Introduction to Structural Dynamics and Aeroelasticity

Aeroelastic and structural dynamic phenomena play an important role in many facets of engineering. In particular, an understanding of these disciplines is essential to the design of aircraft and space vehicles.

This text provides an introduction to structural dynamics and aeroelasticity, with an emphasis on conventional aircraft. The primary areas considered are structural dynamics, static aeroelasticity, and dynamic aeroelasticity. The structural dynamics material emphasizes vibration, the modal representation, and dynamic response. Aeroelastic phenomena discussed include divergence, aileron reversal, airload redistribution, unsteady aerodynamics, flutter, and elastic tailoring. Both exact and approximate solution methodologies are stressed. More than 100 illustrations and tables help clarify the text, while upwards of 50 problems enhance student learning.

This text meets the need for an up-to-date treatment of structural dynamics and aeroelasticity for advanced undergraduate or beginning graduate aerospace engineering students.

Dewey H. Hodges is Professor in the School of Aerospace Engineering at the Georgia Institute of Technology.

G. Alvin Pierce is Professor Emeritus in the School of Aerospace Engineering at the Georgia Institute of Technology.

Cambridge Aerospace Series

General Editors
MICHAEL J. RYCROFT AND WEI SHYY

Introduction to Structural Dynamics and Aeroelasticity

DEWEY H. HODGES
Georgia Institute of Technology

G. ALVIN PIERCE
Georgia Institute of Technology

CAMBRIDGE
UNIVERSITY PRESS

CAMBRIDGE UNIVERSITY PRESS
Cambridge, New York, Melbourne, Madrid, Cape Town, Singapore, São Paulo

Cambridge University Press
32 Avenue of the Americas, New York, NY 10013-2473, USA

www.cambridge.org
Information on this title: www.cambridge.org/9780521806985

First published 2002
Reprinted 2004, 2005, 2006

Printed in the United States of America

A catalog record for this publication is available from the British Library.

Library of Congress Cataloging in Publication Data

Hodges, Dewey H.
Introduction to structural dynamics and aeroelasticity / Dewey H. Hodges, G. Alvin Pierce.
    p.   cm. – (Cambridge aerospace series ; 15)
Includes bibliographical references and index.
ISBN 0-521-80698-4
1. Airframes – Design and construction.   2. Structural design.   3. Space
vehicles – Dynamics.   4. Aeroelasticity.   I. Pierce, G. Alvin.   II. Title.   III. Series.
TL671.6.H565   2002
629.134′31 – dc21   2001052552

ISBN-13   978-0-521-80698-5 hardback
ISBN-10   0-521-80698-4 hardback

To our wives, Margaret and Kathy

Contents

Foreword

A senior-level undergraduate course entitled "Vibration and Flutter" was taught for many years at Georgia Tech under the quarter system. This course dealt with elementary topics involving the static and/or dynamic behavior of structural elements, both without and with the influence of a flowing fluid. The course did not deal with the static behavior of structures in the absence of fluid flow, because this is typically considered in courses in structural mechanics. Thus, the course essentially dealt with the fields of "Structural Dynamics" (when fluid flow is not considered) and "Aeroelasticity" (when it is).

As the name suggests, structural dynamics is concerned with the vibration and dynamic response of structural elements. It can be regarded as a subset of aeroelasticity, the field of study concerned with interaction between the deformation of an elastic structure in an airstream and the resulting aerodynamic force. Aeroelastic phenomena can be observed on a daily basis in nature (e.g., the swaying of trees in the wind, the humming sound Venetian blinds make in the wind, etc.). The most general aeroelastic phenomena include dynamics, but static aeroelastic phenomena are also quite important. The course has been expanded to cover a full semester, and the course title has been appropriately changed to "Introduction to Structural Dynamics and Aeroelasticity."

Aeroelastic and structural dynamic phenomena can result in dangerous static and dynamic deformations and instabilities and, thus, have very important practical consequences in many areas of technology. Especially when one is concerned with the design of modern aircraft and space vehicles, both of which are characterized by the demand for extremely lightweight structures, the solution of many structural dynamics and aeroelasticity problems is a basic requirement for achieving an operationally reliable and structurally optimal system. Aeroelastic phenomena can also play an important role in turbomachinery, civil engineering structures, wind energy converters, and even in the sound generation of musical instruments.

Aeroelastic problems may be roughly classified into the categories of response and stability. Although stability problems are the principal focus of the material presented herein, this is not because response problems are any less important. Rather, because the amplitude of deformation is indeterminate in linear stability problems, one may consider an exclusively linear treatment and still manage to solve many practical problems. However, because the amplitude is important in response problems, one is far more likely to need to be concerned with nonlinear behavior when attempting to solve them. Although nonlinear equations come closer to representing reality, analytical solution of nonlinear equations is problematic, especially in the context of undergraduate studies.

The purpose of this text is to provide an introduction to the fields of structural dynamics and aeroelasticity. The length and scope of the text are intended to be appropriate for a semester-length, senior-level, undergraduate course or a first-year graduate course in which the emphasis is placed on conventional aircraft. For curricula that provide a separate course in structural dynamics, an ample amount of material has been added to the aeroelasticity chapters so that a full course on aeroelasticity alone could be developed from this text.

    This text has been built on the foundation provided by Prof. Pierce's course notes, which had been used for the Vibration and Flutter course since the 1970s. After Prof. Pierce's retirement in 1995, when the responsibility for the course was transferred to Prof. Hodges, the idea was conceived of turning the notes into a more substantial text. This process began with the laborious conversion of Prof. Pierce's original set of course notes to LaTeX 2_ε format in the fall of 1997. The authors are grateful to Margaret Ojala, who was at that time Prof. Hodges's administrative assistant and who facilitated that conversion. Prof. Hodges then began the process of expanding the material and adding problems to all chapters. Some of the most substantial additions were in the aeroelasticity chapters, partly motivated by Georgia Tech's conversion to semesters. Dr. Mayuresh J. Patil, while he was a Post-Doctoral Fellow in the School of Aerospace Engineering, worked with Prof. Hodges to add material on aeroelastic tailoring and unsteady aerodynamics mainly during academic year 1999–2000. The authors thank Prof. David A. Peters of Washington University for his comments on the unsteady aerodynamics section. Finally, Prof. Pierce, while enjoying his retirement and building a new house and amidst a computer hardware failure and visits from grandchildren, still managed to add material on the history of aeroelasticity and on the k and $p–k$ methods in the early summer of 2001.

CHAPTER 1

Introduction

Aeroelasticity is the term used to denote the field of study concerned with the interaction between the deformation of an elastic structure in an airstream and the resulting aerodynamic force. The interdisciplinary nature of the field can be best illustrated by Fig. 1.1, which depicts the interaction of the three disciplines of aerodynamics, dynamics, and elasticity. Classical aerodynamic theories provide a prediction of the forces acting on a body of a given shape. Elasticity provides a prediction of the shape of an elastic body under a given load. Dynamics introduces the effects of inertial forces. With the knowledge of elementary aerodynamics, dynamics, and elasticity, the student is in a position to look at problems in which two or more of these phenomena interact. One of those areas of interaction is the field of flight mechanics, which most students have studied in a separate course by the senior year. The present text will consider the three remaining areas of interaction:

- between elasticity and dynamics (structural dynamics),
- between aerodynamics and elasticity (static aeroelasticity), and
- among all three (dynamic aeroelasticity).

Because of their importance to aerospace system design, these are appropriate for study in an undergraduate aeronautics/aeronautical engineering curriculum. In aeroelasticity one finds that the loads depend on the deformation (aerodynamics), and the deformation depends on the loads (structural mechanics/dynamics); thus one has a coupled problem. Consequently, prior study of all three constituent disciplines is necessary before a study in aeroelasticity can be undertaken. Moreover, a study in structural dynamics can be helpful to develop concepts that are useful in solving aeroelasticity problems, such as the modal representation.

It is of interest to note that aeroelastic phenomena have played a major role throughout the history of powered flight. The Wright brothers utilized controlled warping of the wings on their Wright Flyer in 1903 to achieve lateral control. This was essential to their success in achieving powered flight because the aircraft was laterally unstable owing to the significant anhedral of the wings. Earlier in 1903 Samuel Langley made two attempts to achieve powered flight from the top of a houseboat on the Potomac River. His efforts resulted in catastrophic failure of the wings caused by their being overly flexible and overloaded. Such aeroelastic phenomena, including torsional divergence, were major factors in the predominance of the biplane design until the early 1930s when "stressed skin" metallic structural configurations were introduced to provide adequate torsional stiffness for monoplanes.

The first recorded and documented case of flutter in an aircraft occurred in 1916. The Handley Page O/400 bomber experienced violent tail oscillations as the result of the lack of a torsion rod connection between the port and starboard elevators, an absolute design requirement of today. The incident involved a dynamic twisting of the fuselage to as much as 45 degrees in conjunction with an antisymmetric flapping of the elevators. Catastrophic failures due to aircraft flutter became a major design concern during the First World War and remain so today. R. A. Frazer and W. J. Duncan at the National Physical Laboratory in England compiled a classic document on this subject entitled "The Flutter of Aeroplane

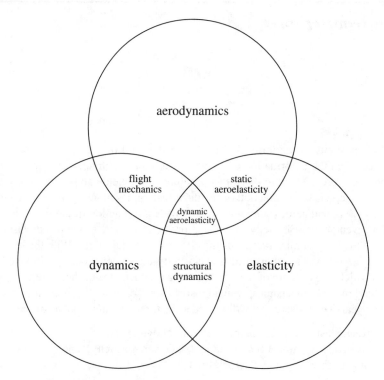

Figure 1.1 Schematic of the field of aeroelasticity.

Wings" as R&M 1155 in August 1928. This small document (about 200 pages) became known as "The Flutter Bible." Their treatment for the analysis and prevention of the flutter problem laid the groundwork for the techniques in use today.

Another major aircraft design concern that may be classified as a static aeroelastic phenomenon was experienced in 1927 by the Bristol Bagshot, a twin-engine, high-aspect-ratio English aircraft. As the speed was increased the aileron effectiveness decreased to zero and then became negative. This loss and reversal of aileron control is commonly known today as aileron reversal. The incident was successfully analyzed and design criteria were developed for its prevention by Roxbee Cox and Pugsley at the Royal Aircraft Establishment in the early 1930s. Although aileron reversal does not generally lead to a catastrophic failure, it can be quite dangerous and is thus an essential design concern. It is of interest to note that during this period of the early 1930s it was Roxbee Cox and Pugsley who proposed the name "aeroelasticity" to describe these phenomena, which are the subject of this text.

In the design of aerospace vehicles, aeroelastic phenomena can result in a full spectrum of behavior from the near benign to the catastrophic. At the near benign end of the spectrum one finds passenger and pilot discomfort. One moves from there to steady-state and transient vibrations that slowly cause the aircraft structure to suffer fatigue damage at the microscopic level. At the catastrophic end, there are aeroelastic instabilities that can quickly destroy an aircraft and result in loss of human life without warning. Aeroelastic problems that need to be addressed by the aerospace system designer can be mainly static in nature, meaning that inertial forces do not play a significant role, or they can be strongly influenced by inertial forces. Although not the case in general, the analysis of some aeroelastic phenomena can be undertaken by means of small deformation theories. Aeroelastic phenomena may strongly affect the performance of an aircraft, positively or negatively. They may also determine whether its control surfaces perform their intended functions well, poorly, or even in the exact opposite manner of that which they are intended to do. It is clear then that all these studies have very important practical consequences in many areas of aerospace technology. The design of modern aircraft and space vehicles is characterized by the demand for extremely lightweight structures. Therefore, the solution of many aeroelastic problems is a basic requirement for achieving an operationally reliable and structurally optimal system. Aeroelastic phenomena also play an important role in turbomachinery, in wind energy converters, and even in the sound generation of musical instruments.

The most commonly posed problems for the aeroelastician are stability problems. Although the elastic moduli of a given structural member are independent of the speed of the aircraft, the aerodynamic forces strongly depend on it. It is therefore not difficult to imagine scenarios in which the aerodynamic forces "overpower" the elastic restoring forces. When this occurs in such a way that inertial forces have little effect, we refer to this as a static aeroelastic instability – or "divergence." In contrast, when the inertial forces are important, the resulting dynamic instability is called "flutter." Both divergence and flutter can be catastrophic, leading to sudden destruction of the vehicle. Thus, it is vital for aircraft designers to know how to design lifting surfaces that are free of such problems. Most of the treatment of aeroelasticity in this text is concerned with stability problems.

Much of the rest of the field of aeroelasticity involves a study of the response of aircraft in flight. Static aeroelastic response problems constitute a special case in which inertial forces do not contribute and in which one may need to predict the lift developed by an aircraft of given configuration at a specified angle of attack, or determine the maximum load factor such an aircraft can sustain. Also, problems of control effectiveness and aileron reversal fall under this category. When inertial forces are important, one may need to know how the aircraft reacts in turbulence or in gusts. Another important phenomenon is buffeting, which is characterized by transient vibration induced by wakes behind wings, nacelles, or other components of the aircraft.

All the above are treatable within the context of a linear analysis. Mathematically, linear aeroelastic response and stability problems are complementary. That is, instabilities are predictable from examining the situations under which homogeneous equations possess nontrivial solutions. Response problems, however, are generally based on solution of nonhomogeneous equations. When the system goes unstable, a solution to the nonhomogeneous equations ceases to exist, while the homogeneous equations and boundary conditions associated with a stable configuration have no nontrivial solution.

Unlike the predictions from linear analyses, in real aircraft it is possible for self-excited oscillations to develop, even at speeds less than the flutter speed. Moreover, large disturbances can bump a system that is predicted to be stable by linear analyses into a state of large oscillatory motion. Both situations can lead to steady-state periodic oscillations for the entire system called limit-cycle oscillations (LCO). In such situations, there can be fatigue problems leading to concerns about the life of certain components of the aircraft as well as passenger comfort and pilot endurance. To capture such behavior in an analysis, the aircraft must be treated as a nonlinear system. Although of great practical importance, nonlinear analyses are beyond the scope of this textbook.

The organization of the text is now presented. To describe the dynamic behavior of conventional aircraft, the topic of *structural dynamics* is introduced in Chapter 2. This is the study of dynamic properties of continuous elastic configurations, which provides a means of analytically representing a flight vehicle's deformed shape at any instant of time. We begin with very simple systems, such as vibrating strings, and move up in complexity to beams in torsion and finally to beams in bending. The introduction of the modal representation and its subsequent use in solving aeroelastic problems is the paramount emphasis of this chapter. A very brief introduction to the methods of Ritz and Galerkin is also included.

Chapter 3 treats *static aeroelasticity*. Therein we concern ourselves with static instabilities, steady airloads, and control effectiveness problems. Again, we begin with simple systems, such as elastically restrained rigid wings. We move up to wings in torsion and finish the chapter with a treatment of swept wings undergoing elastically coupled bending and torsion deformation. Finally, Chapter 4 treats *aeroelastic flutter*, which is associated with dynamic aeroelastic instabilities due to the mutual interaction of aerodynamic, elastic and inertial forces. A generic lifting surface analysis is first presented, and this is followed by illustrative treatments involving simple "typical section models." Engineering solution methods for flutter are discussed, followed by a brief presentation of unsteady aerodynamic theories, both classical and modern. The chapter culminates with an application of the modal representation to the flutter analysis of flexible wings and a discussion of the flutter boundary characteristics of conventional aircraft. It is important to note that central to our study in these last two chapters are the phenomena of divergence and flutter, which typically result in catastrophic failure of the lifting surface and may lead to subsequent destruction of the flight vehicle.

Lists of references for structural dynamics and aeroelasticity are included, along with an appendix in which Lagrange's equations are derived and illustrated.

CHAPTER 2

Structural Dynamics

O students, study mathematics, and do not build without foundations....

Leonardo da Vinci

The purpose of this chapter is to convey to the student a small, introductory portion of the theory of structural dynamics. Much of the theory to which the student will be exposed in this treatment was developed by mathematicians during the time between Newton and Rayleigh. The grasp of this mathematical foundation is therefore a goal that is worthwhile in its own right. Moreover, as implied by the above quotation, a proper use of this foundation enables the advance of technology.

The field of structural dynamics addresses the dynamic deformation behavior of continuous structural configurations. In general, load–deflection relationships are nonlinear, and the deflections are not necessarily small. In this chapter, to facilitate tractable, analytical solutions, we restrict our attention to linearly elastic systems undergoing small deflections, conditions that typify most flight vehicle operations. It should be noted, however, that some level of geometrically nonlinear theory is necessary to arrive at a set of linear equations for strings, membranes, helicopter blades, turbine blades, and flexible rods in rotating spacecraft. Among these problems, only strings are treated herein. Indeed, linear equations of motion for free vibration of strings cannot be obtained without initial consideration, and subsequent careful elimination, of nonlinearities. Finally, there are other important phenomena, such as limit-cycle oscillations in lifting surfaces, that must be treated with sophisticated nonlinear analysis methodology; but they are beyond the scope of this text.

Structural dynamics is a broad subject, covering such things as determination of natural frequencies and mode shapes (the so-called free-vibration problem), response due to initial conditions, forced response in the time domain, and frequency response. In the following we will deal with all except the last category. For response problems, if the loading is at least in part of aerodynamic origin, then the response is said to be aeroelastic. In general the aerodynamic loading will then depend on the structural deformation, and the deformation will depend on the aerodynamic loading. As with the structural dynamics, the aerodynamic forces driving the system may be nonlinear. Linear aeroelastic problems are considered in later chapters.

The value of structural dynamics to the general study of aeroelastic phenomena is its ability to provide a means of quantitatively describing the deformation pattern at any instant in time for a continuous structural system in response to external loading. Although there are many methods of approximating the structural deformation pattern, several of the widely used methods are reducible to what is called a modal representation as long as the underlying structural modeling is linear. It is the purpose of this chapter to establish the concept of modal representation and show how it can be used to describe the dynamic behavior of continuous elastic systems. Also included is an introductory treatment of the Ritz and Galerkin methods, techniques that make use of mode shapes or a similar set of functions to obtain approximate solutions in a simple way. Both methods are indeed close relatives of the finite element method, a widely used approximate method that can accurately analyze realistic structural configurations. The finite element method is not covered herein, but details of this method can be found in books that offer a more advanced perspective on structural analysis, several of which are listed in the bibliography.

The analytical developments presented in this chapter are conceptually similar to the methods of analysis conducted on complete flight vehicles. In an effort to maintain analytical simplicity, the continuous structural configurations to be examined are all uniform and one-dimensional. Although such structures may appear impractical in relation to conventional aircraft, they exhibit structural dynamic properties and representations that are essentially the same as those of full-scale flight vehicles.

2.1   Uniform String Dynamics

To more easily understand a mathematical description of the mechanics associated with the structural dynamics of continuous elastic systems, the classical "vibrating string problem" will first be considered. Although the string can be described by a simple second-order partial differential equation in one dimension, it is typically descriptive of the more complex linearly elastic systems of aerospace vehicles. Once the fundamental concepts are covered for the string, other components will be treated that are more representative of these vehicles. Although the free vibration of a string can be analyzed using equations of motion that are of the same form as that of uniform beam extensional and torsional vibrations, the string is chosen as our first example primarily because, in contrast to these other structures, string behavior can be so easily visualized. Moreover, by this time in their undergraduate studies, most students have had some exposure to the solution of string vibration problems.

2.1.1   *Equations of Motion*

A uniform string of initial length ℓ_0 will be stretched in the x-direction between two walls separated by a distance $\ell > \ell_0$. The string tension T will be considered high and the transverse displacements $v(x, t)$ will eventually be regarded as small. At any given instant this system can be illustrated as in Fig. 2.1. To describe the dynamic behavior of this system the forces acting on a differential length dx of the string can be illustrated by Fig. 2.2. It may be noted that the tension and slope, θ, at the right end of the differential element have been represented as a Taylor series expansion of the values at the left end. Because the string segment is of a differential length that can be arbitrarily small, the series will be truncated by neglecting terms of the order of dx^2 and higher.

Two equations of motion can be formed by resolving the tension forces in the x- and y-directions and setting the resultant force on the differential element equal to its mass

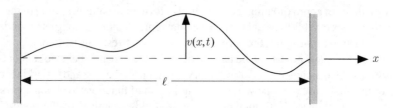

Figure 2.1   Schematic of vibrating string.

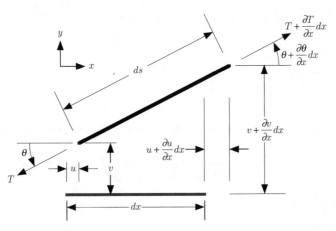

Figure 2.2 Differential element of string showing displacement components and tension force.

$(m\,dx)$ times the acceleration of its mass center. Thus, the equations of motion are

$$\frac{\partial}{\partial x}\left[T\cos(\theta)\right] = m\frac{\partial^2 u}{\partial t^2},$$

$$\frac{\partial}{\partial x}\left[T\sin(\theta)\right] = m\frac{\partial^2 v}{\partial t^2}, \tag{2.1}$$

where $m = \rho A$ is the mass per unit length. From the picture, ignoring second and higher powers of dx and letting $ds = (1 + e)\,dx$ where e is the elongation, one can identify

$$\cos(\theta) = \frac{1}{1+e}\left(1 + \frac{\partial u}{\partial x}\right),$$

$$\sin(\theta) = \frac{1}{1+e}\frac{\partial v}{\partial x}. \tag{2.2}$$

Noting that $\cos^2(\theta) + \sin^2(\theta) = 1$, one can find the elongation e as

$$e = \frac{\partial s}{\partial x} - 1 = \sqrt{\left(1 + \frac{\partial u}{\partial x}\right)^2 + \left(\frac{\partial v}{\partial x}\right)^2} - 1. \tag{2.3}$$

Finally, considering the string as linearly elastic, one can write the tension force as a linear function of the elongation, so that

$$T = EAe, \tag{2.4}$$

where EA is the constant longitudinal stiffness of the string. This completes the system of nonlinear equations that govern the vibration of the string. In order for us to develop analytical solutions, these equations must be simplified.

Let us presuppose the existence of a static equilibrium solution of the string deflection so that

$$u(x, t) = \bar{u}(x),$$

$$v(x, t) = 0,$$

$$\theta(x, t) = 0, \tag{2.5}$$

$$e(x, t) = \bar{e}(x),$$

$$T(x, t) = \bar{T}(x).$$

One then finds that such a solution exists and that, if $\overline{u}(0) = 0$,

$$\overline{T}(x) = T_0,$$

$$\overline{e}(x) = e_0 = \frac{T_0}{EA} = \frac{\delta}{\ell_0}, \tag{2.6}$$

$$\overline{u}(x) = e_0 x,$$

where $\delta = \ell - \ell_0$ is the change in length of the string between its stretched and unstretched states.

If the steady-state tension T_0 is sufficiently high, the perturbation deflections about the static equilibrium solution are very small. Thus, we can assume

$$u(x, t) = \overline{u}(x) + \hat{u}(x, t),$$

$$v(x, t) = \hat{v}(x, t),$$

$$\theta(x, t) = \hat{\theta}(x, t), \tag{2.7}$$

$$e(x, t) = \overline{e}(x) + \hat{e}(x, t),$$

$$T(x, t) = \overline{T}(x) + \hat{T}(x, t),$$

where the $(\hat{\ })$-quantities are taken to be infinitesimally small. Furthermore, from the second of Eqs. (2.2) one can determine $\hat{\theta}$ in terms of the other quantities, that is,

$$\hat{\theta} = \frac{1}{1 + e_0} \frac{\partial \hat{v}}{\partial x}. \tag{2.8}$$

Substituting the perturbation expressions of Eqs. (2.7) and (2.8) into Eqs. (2.1) while ignoring all squares and products of the $(\hat{\ })$-quantities, one finds that the equations of motion can be reduced to two linear, partial differential equations:

$$EA \frac{\partial^2 \hat{u}}{\partial x^2} = m \frac{\partial^2 \hat{u}}{\partial t^2},$$

$$\frac{T_0}{1 + e_0} \frac{\partial^2 \hat{v}}{\partial x^2} = m \frac{\partial^2 \hat{v}}{\partial t^2}. \tag{2.9}$$

Thus, the two nonlinear equations of motion, Eqs. (2.1), for the free vibration of a string have been reduced to two uncoupled linear equations, one for longitudinal vibration and the other for transverse vibration. Since it is typically true that $EA \gg T_0$, longitudinal motions have much smaller amplitudes and much higher natural frequencies; thus, they are not usually of interest. Moreover, the fact that $EA \gg T_0$ also leads to the observations that $e_0 \ll 1$ and $\delta \ll \ell_0$ [see Eqs. (2.6)]. Thus, the transverse motion is governed by

$$T_0 \frac{\partial^2 \hat{v}}{\partial x^2} = m \frac{\partial^2 \hat{v}}{\partial t^2}. \tag{2.10}$$

For convenience, we will drop the $(\hat{\ })$s and the subscript, thus yielding the usual equation for string vibration found in texts on vibration:

$$T \frac{\partial^2 v}{\partial x^2} = m \frac{\partial^2 v}{\partial t^2}. \tag{2.11}$$

This is called the one-dimensional "wave equation" and governs the structural dynamic behavior of the string in conjunction with the boundary conditions. These conditions at the ends of the string correspond to zero displacement as described by

$$v(0, t) = v(\ell, t) = 0, \tag{2.12}$$

where it is noted that the distinction between ℓ_0 and ℓ is no longer relevant. The general solution to the above wave equation with these homogeneous boundary conditions comprises a simple eigenvalue problem. The fact that the equation is of second order both temporally and spatially indicates that two boundary conditions and two initial conditions need to be specified.

2.1.2 *Standing Wave (Modal) Solution*

The preceding wave equation, which governs the dynamic behavior of the string, can be reduced from a partial differential equation with two independent variables to two ordinary differential equations by making a "separation of variables." The dependent variable of transverse displacement will be represented by

$$v(x, t) = X(x)Y(t). \tag{2.13}$$

This product form will now be substituted into the wave equation, Eq. (2.11). To simplify the notation let $(\ )'$ and $(\dot{\ })$ denote ordinary derivatives with respect to x and t. Thus, the wave equation becomes

$$T X''(x)Y(t) = m X(x)\ddot{Y}(t). \tag{2.14}$$

Rearranging terms as

$$\frac{X''(x)}{X(x)} = \frac{m\ddot{Y}(t)}{TY(t)}. \tag{2.15}$$

one observes that the left-hand side of this equation is only a function of the single independent variable x and the right-hand side is a function of only t. This, of course, presumes that both m and T are constants. Constant m implies that the string is uniform, and constant T is consistent with the prior approximations used in deriving Eq. (2.11). Since each side of the equation is a function of different independent variables, then the only way the equality can be valid is for each side to be equal to a common constant. Let this constant be $-\alpha^2$, so that

$$\frac{X''(x)}{X(x)} = \frac{m\ddot{Y}(t)}{TY(t)} = -\alpha^2. \tag{2.16}$$

This yields two ordinary differential equations, given by

$$X''(x) + \alpha^2 X(x) = 0,$$
$$\ddot{Y}(t) + \alpha^2 \frac{T}{m} Y(t) = 0. \tag{2.17}$$

Because the general solutions to these linear, second-order, differential equations are well known they will be written without any further justification as

$$X(x) = A \sin(\alpha x) + B \cos(\alpha x),$$
$$Y(t) = C \sin\left(\sqrt{\frac{T}{m}}\alpha t\right) + D \cos\left(\sqrt{\frac{T}{m}}\alpha t\right). \tag{2.18}$$

These solutions are only valid when $\alpha \neq 0$. The much simpler solutions for the special case of $\alpha = 0$ will be discussed later. The boundary condition on the left end of the string, where $x = 0$, can be written as

$$v(0, t) = X(0)Y(t) = 0, \tag{2.19}$$

which is satisfied when

$$X(0) = 0, \tag{2.20}$$

and so

$$B = 0. \tag{2.21}$$

The boundary condition on the right end is

$$v(\ell, t) = X(\ell)Y(t) = 0, \tag{2.22}$$

which is satisfied when

$$X(\ell) = 0, \tag{2.23}$$

and so

$$A \sin(\alpha \ell) = 0. \tag{2.24}$$

If $A = 0$ the displacement will be identically zero for all x and t. Although this is an acceptable solution, it is of little interest and is thus called a trivial solution. Of more concern is when

$$\sin(\alpha \ell) = 0. \tag{2.25}$$

This relation is called the "characteristic equation" and has a denumerably infinite set of solutions known as "eigenvalues." These solutions can be written as

$$\alpha_i = \frac{i\pi}{\ell} \qquad (i = 1, 2, \ldots). \tag{2.26}$$

It should be noted that, although $i = 0$ (implying $\alpha_0 = 0$) appears to lead to a trivial solution, this is not true in general. Although $\alpha = 0$ does lead to a trivial solution in this case, the only way to ascertain whether a nontrivial $\alpha = 0$ solution exists in the general case is to return to Eqs. (2.17) and determine whether an $\alpha = 0$ solution to those equations satisfies all the boundary conditions. Obviously, it does not here. Additional examples associated with the $\alpha = 0$ solution will be addressed in more detail later when we consider problems other than strings.

Therefore, for each integer value of the index i there is an eigenvalue α_i and an associated solution X_i, called the "eigenfunction." It contributes to the general solution based on the corresponding value of Y_i. Thus, its total contribution can be written as

$$v_i(x, t) = X_i(x)Y_i(t), \tag{2.27}$$

where

$$X_i(x) = A_i \sin(\alpha_i x),$$

$$Y_i(t) = C_i \sin\left(\sqrt{\frac{T}{m}}\alpha_i t\right) + D_i \cos\left(\sqrt{\frac{T}{m}}\alpha_i t\right). \tag{2.28}$$

Note that the constants A_i, C_i, and D_i may have different numerical values for each eigenvalue; thus they have been subscripted with the index. The most general solution for the string displacement would have contributions associated with all the eigenvalues. Thus, the general solution can be written as a sum of the complete set as

$$
\begin{aligned}
v(x, t) &= \sum_{i=1}^{\infty} v_i(x, t) \\
&= \sum_{i=1}^{\infty} \sin\left(\frac{i\pi x}{\ell}\right)\left[E_i \sin\left(\sqrt{\frac{T}{m}}\frac{i\pi t}{\ell}\right) + F_i \cos\left(\sqrt{\frac{T}{m}}\frac{i\pi t}{\ell}\right)\right].
\end{aligned}
\tag{2.29}
$$

Note that the original constants have been combined as

$$
\begin{aligned}
E_i &= A_i C_i, \\
F_i &= A_i D_i.
\end{aligned}
\tag{2.30}
$$

Close inspection of this total string displacement indicates that at any given instant the transverse deflection is represented by summation over a denumerably infinite set of shapes. Each shape is of indeterminate amplitude and is associated with a particular eigenfunction; these shapes are also called "mode shapes" in the field of structural dynamics. They will be represented here by $\phi_i(x)$. Thus, for transverse deflection of a string the mode shapes may be written as

$$
\phi_i(x) = \sin\left(\frac{i\pi x}{\ell}\right)
\tag{2.31}
$$

or any constant times $\phi_i(x)$. It can be observed from this function (see Fig. 2.3) that the higher the mode number i, the more crossings of the zero axis on the interval $0 < x < \ell$. These crossings are sometimes referred to as "nodes." The trend of increasing numbers of nodes with an increase in the mode number is generally true in structural dynamics.

In the above solution for the total displacement it may be noted that each mode shape is multiplied by a function of time. This multiplier is called the "generalized coordinate" and will be represented here by $\xi_i(t)$. For this specific problem the generalized coordinates are

$$
\xi_i(t) = E_i \sin\left(\sqrt{\frac{T}{m}}\frac{i\pi t}{\ell}\right) + F_i \cos\left(\sqrt{\frac{T}{m}}\frac{i\pi t}{\ell}\right)
\tag{2.32}
$$

and are thus seen to be simple harmonic functions of time. Since there were no external loads applied to the string, the preceding result is called the homogeneous solution. If there had been an external loading, the resulting time dependency of the generalized coordinates would reflect such a loading.

Thus, the total string displacement can be written as a sum of "modal" contributions of the form

$$
v(x, t) = \sum_{i=1}^{\infty} \phi_i(x)\xi_i(t).
\tag{2.33}
$$

This expression can be interpreted as a weighted sum of the mode shapes, each of which has a modal amplitude (i.e., the generalized coordinate) that is a function of time. For the homogeneous solution obtained above, this time dependency is simple harmonic at a frequency that is unique for each mode or eigenvalue. These are called the "natural frequencies" of the modes or "modal frequencies" and will be represented by ω_i. For the string they are

$$
\omega_i = \frac{i\pi}{\ell}\sqrt{\frac{T}{m}} \qquad (i = 1, 2, \ldots, \infty),
\tag{2.34}
$$

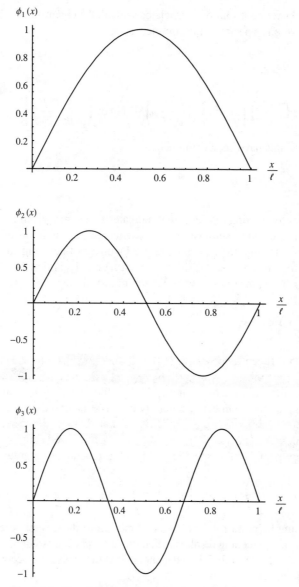

Figure 2.3 First three mode shapes for vibrating string.

with the lowest frequencies given by the lowest mode numbers. Indeed, just as the increase in the number of nodes with the mode number is generally true, so it is with the natural frequency. When the physical and geometric parameters of the problem are expressed in any consistent[1] set of units, the units of the natural frequency will be rad/s. Division by 2π converts the units of frequency into "cycles per second" or Hertz. The inverse of the natural frequency in Hertz is the period of the oscillatory motion.

---

[1] For example, with SI units one has the units of T as N, m as kg/m, and ℓ as m. With English units, one has the units of T as lb, m as, say, lb-s^2/in.^{2}, and ℓ as in.

To summarize what has been accomplished in solving the wave equation, it may be said that the string displacement as a function of both x and t can be represented as a sum of modal contributions. Each mode in this representation is a structural dynamic property of the given system (string) and can be completely described by its mode shape and modal frequency. Such "modes of vibration" can be formulated for any conservatively loaded, linearly elastic structure. This statement includes two restrictions that must be observed for a modal representation. One restriction is linearity, which is satisfied here by the linear wave equation. The other restriction is that the system must be conservative, which means that there can be no addition or dissipation of energy during the dynamic response. A typical violation of this restriction would be the existence of damping such as structural or aerodynamic damping. When such damping is present, it can be adequately treated as an external loading.

2.1.3   Orthogonality of Mode Shapes

A most significant property of the mode shapes previously discussed is that they form a set of orthogonal mathematical functions. If the system is nonuniform then the mode shapes are orthogonal with respect to some inertial weighting function, such as the mass distribution for the string. This condition of functional orthogonality can be described analytically as

$$\int_0^\ell m\phi_i(x)\phi_j(x)\,dx = 0 \qquad (i \neq j),$$
$$\neq 0 \qquad (i = j). \qquad (2.35)$$

To prove that the mode shapes obtained for the uniform string problem are orthogonal, an individual modal contribution given by

$$v_i(x,t) = \phi_i(x)\xi_i(t) \qquad (2.36)$$

is substituted into the governing differential equation (wave equation) to obtain

$$T\frac{\partial^2 v_i}{\partial x^2} = m\frac{\partial^2 v_i}{\partial t^2}. \qquad (2.37)$$

Because the generalized coordinate is a simple harmonic function for the homogeneous solution,

$$\ddot{\xi}_i = -\omega_i^2\,\xi_i, \qquad (2.38)$$

and the wave equation becomes

$$T\phi_i''(x)\xi_i(t) = -m\phi_i(x)\omega_i^2\xi_i(t), \qquad (2.39)$$

so that

$$T\phi_i''(x) = -m\phi_i(x)\omega_i^2. \qquad (2.40)$$

If this procedure is repeated by substituting the jth modal contribution into the wave equation, a similar result,

$$T\phi_j''(x) = -m\phi_j(x)\omega_j^2, \qquad (2.41)$$

is obtained. After multiplying Eq. (2.40) by ϕ_j and Eq. (2.41) by ϕ_i, subtracting, and integrating the result over the length of the string, we obtain

$$(\omega_i^2 - \omega_j^2) m \int_0^\ell \phi_i(x)\phi_j(x)\,dx = T \int_0^\ell [\phi_i(x)\phi_j''(x) - \phi_i''(x)\phi_j(x)]\,dx. \qquad (2.42)$$

The integral on the right-hand side can be integrated by parts using

$$\int_a^b u\,dv = uv \Big|_a^b - \int_a^b v\,du \qquad (2.43)$$

by letting

$$\begin{aligned} u &= \phi_i, & du &= \phi_i'\,dx, \\ v &= \phi_j', & dv &= \phi_j''\,dx \end{aligned} \qquad (2.44)$$

for the first term and

$$\begin{aligned} u &= \phi_j, & du &= \phi_j'\,dx, \\ v &= \phi_i', & dv &= \phi_i''\,dx \end{aligned} \qquad (2.45)$$

for the second term. The result becomes

$$\begin{aligned} (\omega_i^2 - \omega_j^2) m \int_0^\ell \phi_i(x)\phi_j(x)\,dx \\ = T(\phi_i\phi_j' - \phi_i'\phi_j)\big|_0^\ell - T \int_0^\ell (\phi_i'\phi_j' - \phi_i'\phi_j')\,dx = 0. \end{aligned} \qquad (2.46)$$

Note that the right-hand side is zero because ϕ_i is zero at both ends by virtue of the original boundary conditions. It may now be concluded that $\omega_i \neq \omega_j$ when $i \neq j$, so that

$$m \int_0^\ell \phi_i(x)\phi_j(x)\,dx = 0. \qquad (2.47)$$

However, when $i = j$

$$m \int_0^\ell \phi_i^2(x)\,dx = M_i. \doteq \underline{\frac{m\ell}{2}} = \underline{\frac{M_s}{2}} \qquad (2.48)$$

This integral value, M_i, is called the "generalized mass" of the ith mode. These relations thus demonstrate that the mode shapes for the string, which is fixed at both ends, form an orthogonal set of functions.

The above development is for a string of constant mass per unit length and constant tension force. It is important to note that it can readily be generalized for nonuniform mass per unit length. In more involved developments for beam bending and torsional deformation, the structural stiffnesses, which are analogous to the tension force in the string problem, may also be nonuniform along the span. Although these quantities may not be taken outside the integrals in such cases, the rest of the development remains quite similar.

2.1.4   Using Orthogonality

The property of orthogonality is very useful in many aspects of structural dynamics analysis. As an illustration, consider the so-called homogeneous initial condition problem for the string. In this case there are no external loads on the string, but it is presumed to have an initial deflection shape and an initial velocity distribution. Let these initial conditions be

$*M_s = m\ell$ (string mass)

Each mode is characterized by ω_i (or f_i), $\phi_i(x)$, $\xi_i(t)$, M_i

represented as

$$v(x, 0) = f(x),$$

$$\frac{\partial v}{\partial t}(x, 0) = g(x),$$

(2.49)

where it should be noted that both $f(x)$ and $g(x)$ must certainly be compatible with the boundary conditions.

Using Eq. (2.29), these initial conditions can be written in terms of modal representation as

$$v(x, 0) = \sum_{i=1}^{\infty} F_i \sin\left(\frac{i\pi x}{\ell}\right) = f(x),$$

$$\frac{\partial v}{\partial t}(x, 0) = \sum_{i=1}^{\infty} \frac{E_i i\pi}{\ell} \sqrt{\frac{T}{m}} \sin\left(\frac{i\pi x}{\ell}\right) = g(x).$$

(2.50)

Both of these relations will be multiplied by $\sin(j\pi x/\ell)$ and integrated over the length of the string, yielding

$$\int_0^\ell f(x)\sin\left(\frac{j\pi x}{\ell}\right) dx = \sum_{i=1}^{\infty} F_i \int_0^\ell \sin\left(\frac{i\pi x}{\ell}\right) \sin\left(\frac{j\pi x}{\ell}\right) dx$$

$$= F_j \int_0^\ell \sin^2\left(\frac{j\pi x}{\ell}\right) dx$$

$$= \frac{F_j \ell}{2}.$$

(2.51)

The above evaluation has made use of the orthogonality property of the mode shapes, which causes every term in the infinite sum to be zero except where $i = j$. The initial velocity relation can be reduced in the same manner, so that

$$\int_0^\ell g(x)\sin\left(\frac{j\pi x}{\ell}\right) dx = \sum_{i=1}^{\infty} \frac{E_i i\pi}{\ell} \sqrt{\frac{T}{m}} \int_0^\ell \sin\left(\frac{i\pi x}{\ell}\right) \sin\left(\frac{j\pi x}{\ell}\right) dx$$

$$= \frac{E_j j\pi}{\ell} \sqrt{\frac{T}{m}} \int_0^\ell \sin^2\left(\frac{j\pi x}{\ell}\right) dx$$

$$= \frac{E_j j\pi}{2} \sqrt{\frac{T}{m}}.$$

(2.52)

This treatment of the initial conditions permits a direct evaluation of the unknown constants (E_i and F_i) in the modal representation of the total string displacement; that is,

$$E_i = \frac{2}{i\pi} \sqrt{\frac{m}{T}} \int_0^\ell g(x)\sin\left(\frac{i\pi x}{\ell}\right) dx,$$

$$F_i = \frac{2}{\ell} \int_0^\ell f(x)\sin\left(\frac{i\pi x}{\ell}\right) dx.$$

(2.53)

Thus, for the prescribed initial conditions given by $f(x)$ and $g(x)$, the resulting string displacement can be described as

$$v(x, t) = \sum_{i=1}^{\infty} \sin\left(\frac{i\pi x}{\ell}\right) [E_i \sin(\omega_i t) + F_i \cos(\omega_i t)].$$

(2.54)

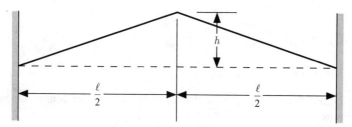

Figure 2.4   Initial shape of plucked string.

Example 1: Response Due to Given Initial Shape

To further illustrate this procedure consider the case of the plucked string with zero initial velocity. Let the initial shape be as shown in Fig. 2.4. If we assume the initial velocity to be zero, then $g(x) = 0$ and $E_i = 0$ for all i. The string displacement becomes

$$v(x, t) = \sum_{i=1}^{\infty} F_i \sin\left(\frac{i\pi x}{\ell}\right) \cos(\omega_i t). \tag{2.55}$$

To evaluate the constants, F_i, the initial string shape will be written as

$$f(x) = \begin{cases} 2h\left(\frac{x}{\ell}\right), & 0 \le x \le \frac{\ell}{2}, \\ 2h\left(1 - \frac{x}{\ell}\right), & \frac{\ell}{2} \le x \le \ell. \end{cases} \tag{2.56}$$

Substituting this function into the preceding integral yields

$$\begin{aligned} F_i &= \frac{4h}{\ell^2}\left[\int_0^{\frac{\ell}{2}} x \sin\left(\frac{i\pi x}{\ell}\right) dx + \int_{\frac{\ell}{2}}^{\ell} (\ell - x) \sin\left(\frac{i\pi x}{\ell}\right) dx\right] \\ &= \frac{8h}{(i\pi)^2} \sin\left(\frac{i\pi}{2}\right). \end{aligned} \tag{2.57}$$

It may be noted that $\sin(i\pi/2)$ is zero for all even values of the index and that it is either positive or negative one for the odd values. If desired these constants can be written as

$$F_i = \begin{cases} \dfrac{8h}{(i\pi)^2}(-1)^{\frac{i-1}{2}} & (i \text{ odd}), \\ 0 & (i \text{ even}). \end{cases} \tag{2.58}$$

The fact that $F_i = 0$ for all even values of i is indicative of the symmetry of the initial string displacement about the midpoint. That is, since the initial shape is symmetric about $x = \ell/2$, no antisymmetric modes of vibration are excited thereby. The total string displacement becomes

$$v(x, t) = \frac{8h}{\pi^2} \sum_{i=1,3,\ldots}^{\infty} \frac{(-1)^{\frac{i-1}{2}}}{i^2} \sin\left(\frac{i\pi x}{\ell}\right) \cos(\omega_i t), \tag{2.59}$$

where

$$\omega_i = \frac{i\pi}{\ell}\sqrt{\frac{T}{m}}. \tag{2.60}$$

It should be noted from this solution that the modal contributions to the total displacement significantly decrease as the mode number (the index i) increases. This can be observed from the dependence of F_i on i. This is characteristic of almost all structural dynamics response problems and thus permits a truncation of the infinite sum to a finite number of the lower-order modes. It also should be noted that the above solution indicates that the string will vibrate forever, with the string returning to its initial shape periodically. In real systems there are always dissipative phenomena that cause the motion to die out in time. This will be considered when we deal with aeroelastic flutter in Chapter 4.

2.1.5   *Traveling Wave Solution*

In the preceding paragraphs a modal solution was obtained for the string problem. This solution depicted the total displacement as a summation of specific shapes as measured relative to the ends of the string. Each shape had an amplitude that was, in general, a function of time. When these individual modal contributions were of constant amplitude at their modal frequency they would appear as standing or fixed waves along the string.

Another interpretation of the string response will now be considered by examining the solution obtained for a string with an initial displacement but zero initial velocity and external loading. In this case the E_is were all zero and so the displacement was written as

$$v(x, t) = \sum_{i=1}^{\infty} \sin\left(\frac{i\pi x}{\ell}\right) F_i \cos\left(\sqrt{\frac{T}{m}} \frac{i\pi t}{\ell}\right). \tag{2.61}$$

The F_is can be determined from the initial shape, $f(x)$, as

$$F_i = \frac{2}{\ell} \int_0^{\ell} f(x) \sin\left(\frac{i\pi x}{\ell}\right) dx. \tag{2.62}$$

It may also be noted that the initial shape can be represented by

$$v(x, 0) = f(x) = \sum_{i=1}^{\infty} F_i \sin\left(\frac{i\pi x}{\ell}\right). \tag{2.63}$$

We note in passing that Eq. (2.63) is known as the Fourier sine series representation of the function $f(x)$. Additional information on Fourier series may be found in more advanced textbooks on structural dynamics and applied mathematics. Now, to rewrite the general solution for this problem, the two well-known identities

$$\sin(\alpha + \beta) = \sin(\alpha)\cos(\beta) + \cos(\alpha)\sin(\beta),$$
$$\sin(\alpha - \beta) = \sin(\alpha)\cos(\beta) - \cos(\alpha)\sin(\beta) \tag{2.64}$$

can be added to yield another identity as

$$\sin(\alpha)\cos(\beta) = \frac{1}{2}[\sin(\alpha + \beta) + \sin(\alpha - \beta)]. \tag{2.65}$$

This identity can be used to rewrite the general solution given by Eq. (2.61) as

$$v(x, t) = \frac{1}{2} \sum_{i=1}^{\infty} F_i \left\{ \sin\left[\frac{i\pi}{\ell}\left(x + \sqrt{\frac{T}{m}}t\right)\right] + \sin\left[\frac{i\pi}{\ell}\left(x - \sqrt{\frac{T}{m}}t\right)\right] \right\}. \tag{2.66}$$

Equation (2.63) gives the functional form of $f(x)$ as an infinite sum of sine functions with coefficients F_i. The two terms on the right-hand side of Eq. (2.66) are of the same form as the sum in Eq. (2.63) and can be identified as having the functional form of $f(x)$ but with different arguments. It is therefore possible to rewrite Eq. (2.66) as

$$v(x,t) = \frac{1}{2}\left[f\left(x + \sqrt{\frac{T}{m}}t\right) + f\left(x - \sqrt{\frac{T}{m}}t\right)\right]. \tag{2.67}$$

This is the principal result of the traveling wave solution. In reality it is mathematically identical to the previously given standing wave solution of Eq. (2.61). The difference is only in one's point of view.

To illustrate how Eq. (2.67) represents traveling waves along the string the two arguments of the shape function will be replaced by new spatial coordinates whose origins are time dependent. These new coordinates are defined as

$$x_L(x,t) \equiv x + \sqrt{\frac{T}{m}}t,$$
$$x_R(x,t) \equiv x - \sqrt{\frac{T}{m}}t. \tag{2.68}$$

Equation (2.67) becomes

$$v(x,t) = \frac{1}{2}[f(x_L) + f(x_R)], \tag{2.69}$$

which indicates that the time-dependent string shape is the sum of two shapes of a form identical to the initial shape but of half its magnitude. Initially at $t = 0$ the origins of the x_L and x_R coincide with the x origin as

$$x_L(x,0) = 0 \quad \text{at} \quad x = 0,$$
$$x_R(x,0) = 0 \quad \text{at} \quad x = 0. \tag{2.70}$$

At any later time $t > 0$, the origins of x_L and x_R can be located by

$$x_L(x,t) = 0 \quad \text{at} \quad x = -\sqrt{\frac{T}{m}}t,$$
$$x_R(x,t) = 0 \quad \text{at} \quad x = \sqrt{\frac{T}{m}}t. \tag{2.71}$$

These results indicate that the x_L coordinate system is moving to the left with a speed $\sqrt{T/m}$ and the x_R coordinate system is moving to the right with the same speed. These origin positions are indicated in Fig. 2.5. As a consequence of these moving origins, the shape $f(x_L)/2$ will appear to propagate to the left and the shape $f(x_R)/2$ will appear

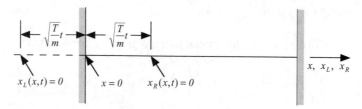

Figure 2.5   Schematic of moving coordinate systems x_L and x_R.

to propagate to the right. Both of these shapes will be moving at a constant propagation speed of

$$V = \sqrt{\frac{T}{m}}, \tag{2.72}$$

so that Eq. (2.67) may be written in the form

$$v(x, t) = \frac{1}{2}[f(x + Vt) + f(x - Vt)]. \tag{2.73}$$

This is also called D'Alembert's form of the equation.

When these shapes reach one of the walls the deflection must go to zero to satisfy the boundary conditions. This condition at each wall causes the shapes to be reflected in the opposite direction. These reflections will appear as inverted shapes propagating away from the walls again with the speed $V = \sqrt{T/m}$. This reflected wave behavior is inherent to the Fourier sine series representation of $f(x)$ given by Eq. (2.63). Determination of the string displacement at times subsequent to $t = 0$ requires the evaluation of $f(x \pm Vt)$ in Eq. (2.67). Although the function $f(x)$ is only defined for the range $0 \leq x \leq \ell$, the arguments $x + Vt$ and $x - Vt$ will significantly exceed this range. The Fourier sine series for $f(x)$, Eq. (2.63), possesses two distinct mathematical properties that permit evaluation of the function throughout the extended range of the argument and demonstrate the reflected wave behavior.

First Property of $f(x)$

Since all terms in the Fourier sine series for $f(x)$ are odd functions of x, then $f(x)$ must also be an odd function. This property can be described as

$$f(-x) = -f(x). \tag{2.74}$$

It is immediately seen that this is a description of the reflected wave behavior at the $x = 0$ wall.

Second Property of $f(x)$

Since all terms in the Fourier sine series for $f(x)$ are periodic in x with a period of 2ℓ, then $f(x)$ must also be periodic in x with a period of 2ℓ. This property can be described as

$$f(x) = f(x + 2n\ell) \qquad \text{for } n = 0, \pm 1, \pm 2, \ldots. \tag{2.75}$$

This relation in conjunction with the previously noted "odd" functionality of $f(x)$ describes the reflected wave behavior at the $x = \ell$ wall.

General Evaluation of $f(x \pm Vt)$

These two properties can be applied simultaneously for the evaluation of $f(x + Vt)$ and $f(x - Vt)$ for any value of their argument, say $x \pm Vt$. When this argument lies within the range

$$n\ell \leq x \pm Vt \leq (n + 1)\ell, \tag{2.76}$$

where

$$n = 0, \pm 1, \pm 2, \ldots, \tag{2.77}$$

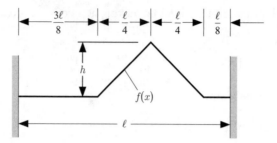

Figure 2.6   Example initial shape of wave.

then

$$f(x \pm Vt) = (-1)^n f \left\{ (-1)^n \left[x \pm Vt + \frac{(-1)^n - 2n - 1}{2} \ell \right] \right\}. \tag{2.78}$$

We have used Eq. (2.75) to reduce the range of motion, which was initially $-\infty \leq x \leq +\infty$, down to the range $0 \leq x \leq \ell$, our physical space (i.e., where the string is actually mounted).

Example 2: Traveling Wave

The initial string shape is given in Fig. 2.6. At subsequent times the string shape will appear as shown in Fig. 2.7. The absolute distance each of the half shapes has traveled at time t is denoted by $\bar{x}$. Note that the faint lines are the displacements associated with the two constituent waves after transformation to bring them into the range $0 \leq x \leq \ell$, while the bold line is the sum of these two displacements. The displacement during the time $\ell\sqrt{m/T} \leq t \leq 2\ell\sqrt{m/T}$ is a mirror image of the progression revealed in Fig. 2.7 with a return to the original shape at $t = 2\ell\sqrt{m/T}$. The motion is periodic thereafter with period $2\ell\sqrt{m/T}$.

2.1.6   Generalized Equations of Motion

Once the free-vibration modes have been determined for a linear, conservative system it is a straightforward procedure to determine the system's response to any external loading. This is accomplished by treating each mode of vibration as a dimensional degree of freedom whose scalar coordinate is the mode's generalized coordinate. For each of these modal degrees of freedom a "generalized equation of motion" can be formulated from "Lagrange's Equations" (see Appendix). Lagrange's equations can be written as

$$\frac{d}{dt}\left(\frac{\partial L}{\partial \dot{\xi}_i}\right) - \frac{\partial L}{\partial \xi_i} = \Xi_i \qquad (i = 1, 2, \ldots), \tag{2.79}$$

where $L = K - P$ is called the "Lagrangean," the difference between the total kinetic energy, K, and the total potential energy, P, of the system. The generalized coordinates are ξ_i; and the term on the right-hand side, Ξ_i, is called the "generalized force." The latter represents the effects of all nonconservative forces, as well as any conservative forces that are not treated in the total potential energy.

Under many circumstances, the kinetic energy can be represented as a function of only the coordinate rates so that

$$K = K(\dot{\xi}_1, \dot{\xi}_2, \dot{\xi}_3, \ldots). \tag{2.80}$$

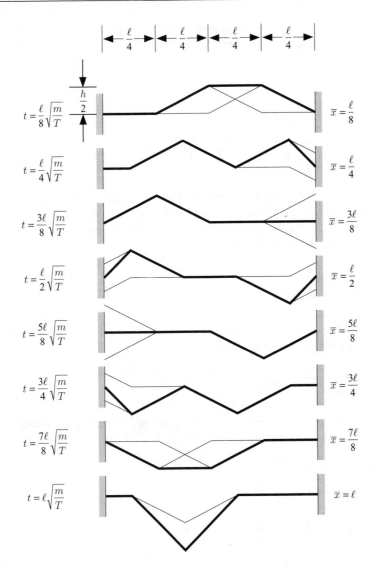

Figure 2.7 Shape of traveling wave at various times.

The potential energy is a function of only the coordinates themselves; that is,

$$P = P(\xi_1, \xi_2, \xi_3, \ldots). \tag{2.81}$$

Thus, Lagrange's equations can be written as

$$\frac{d}{dt}\left(\frac{\partial K}{\partial \dot{\xi}_i}\right) + \frac{\partial P}{\partial \xi_i} = \Xi_i \qquad (i = 1, 2, \ldots). \tag{2.82}$$

The generalized equations of motion for the string problem can be formulated by representing the string displacement in terms of its generalized coordinates as

$$v(x, t) = \sum_{i=1}^{\infty} \phi_i(x)\xi_i(t). \tag{2.83}$$

Because gravitational effects are being neglected, the potential energy of the string will consist of only strain energy caused by extension of the string. This can be expressed as

$$P = \frac{1}{2} \int_0^{\ell_0} EAe^2 dx, \tag{2.84}$$

where, as before,

$$e = \frac{\partial s}{\partial x} - 1 = \sqrt{\left(1 + \frac{\partial u}{\partial x}\right)^2 + \left(\frac{\partial v}{\partial x}\right)^2} - 1 \tag{2.85}$$

and the original length is ℓ_0. In order to pick up all of the linear terms in the generalized equations of motion, one must include all terms in the energy up through the second power of the unknowns. Taking the pertinent unknowns to be perturbations relative to the stretched but undeflected string, we can again write

$$
\begin{aligned}
e(x, t) &= \overline{e}(x) + \hat{e}(x, t), \\
u(x, t) &= \overline{u}(x) + \hat{u}(x, t), \\
v(x, t) &= \hat{v}(x, t).
\end{aligned}
\tag{2.86}
$$

For EA equal to a constant, the strain energy is

$$P = \frac{EA}{2} \int_0^{\ell_0} (\overline{e}^2 + 2\overline{e}\hat{e} + \hat{e}^2) \, dx. \tag{2.87}$$

From Eqs. (2.6), we know that $\overline{T} = T_0$ and $\overline{e} = e_0$, where T_0 and e_0 are constants. Thus, the first term of P is a constant and can be ignored. Since $T_0 = EAe_0$, the strain energy simplifies to

$$P = T_0 \int_0^{\ell_0} \hat{e} \, dx + \frac{EA}{2} \int_0^{\ell_0} \hat{e}^2 \, dx. \tag{2.88}$$

Making use of Eqs. (2.86), one finds that the longitudinal strain becomes

$$\hat{e} = \frac{\partial \hat{u}}{\partial x} + \frac{1}{2(1 + \overline{e})} \left(\frac{\partial \hat{v}}{\partial x}\right)^2 + \dots, \tag{2.89}$$

where the ellipsis refers to terms of third and higher degree in the spatial partial derivatives of $\hat{u}$ and $\hat{v}$. Then, when all terms are dropped that are of third and higher degree in the spatial partial derivatives of $\hat{u}$ and $\hat{v}$, the strain energy becomes

$$P = T_0 \int_0^{\ell_0} \frac{\partial \hat{u}}{\partial x} dx + \frac{T_0}{2(1 + e_0)} \int_0^{\ell_0} \left(\frac{\partial \hat{v}}{\partial x}\right)^2 dx + \frac{EA}{2} \int_0^{\ell_0} \left(\frac{\partial \hat{u}}{\partial x}\right)^2 dx + \dots . \tag{2.90}$$

Assuming $\hat{u}(0) = \hat{u}(\ell_0) = 0$, one finds that the first term vanishes. Since perturbations of the transverse deflections are the unknowns in which we are most interested, and since perturbations of the longitudinal displacements are uncoupled from these and involve oscillations with much higher frequency, we will not need the last term. This leaves only the second term. As before, noting that $e_0 \ll 1$ and dropping the $\hat{\ }$ and subscripts for convenience, one obtains the potential energy for a vibrating string,

$$P = \frac{T}{2} \int_0^\ell \left(\frac{\partial v}{\partial x} \right)^2 dx, \tag{2.91}$$

as found in vibration texts. In terms of the mode shapes as represented in Eq. (2.83) the total potential energy can then be written as

$$P = \frac{T}{2} \int_0^\ell \left(\sum_{i=1}^\infty \phi_i' \xi_i \right)^2 dx. \tag{2.92}$$

Before evaluating this integral it should be noted that the square of a sum (as appearing in the integrand) can be written in terms of a double sum. This can be demonstrated by the following simple example:

$$
\begin{aligned}
\left(\sum_{i=1}^3 a_i \right)^2 &= (a_1 + a_2 + a_3)^2 \\
&= a_1^2 + a_2^2 + a_3^2 + 2a_1 a_2 + 2a_2 a_3 + 2a_3 a_1 \\
&= a_1 (a_1 + a_2 + a_3) + a_2 (a_1 + a_2 + a_3) + a_3 (a_1 + a_2 + a_3) \\
&= a_1 \sum_{i=1}^3 a_i + a_2 \sum_{i=1}^3 a_i + a_3 \sum_{i=1}^3 a_i \\
&= \sum_{j=1}^3 a_j \sum_{i=1}^3 a_i = \sum_{i=1}^3 \sum_{j=1}^3 a_i a_j.
\end{aligned}
\tag{2.93}
$$

Thus, the potential energy becomes

$$P = \frac{T}{2} \sum_{i=1}^\infty \sum_{j=1}^\infty \xi_i \xi_j \int_0^\ell \phi_i' \phi_j' dx. \tag{2.94}$$

For the string, the mode shapes and their first derivatives are sinusoidal functions; consequently, they form an orthogonal set.[2] That is,

$$\int_0^\ell \phi_i'(x) \phi_j'(x)\, dx = 0 \qquad (i \neq j). \tag{2.95}$$

Thus, the potential energy relation can be simplified to

$$P = \frac{T}{2} \sum_{i=1}^\infty \xi_i^2 \int_0^\ell \phi_i'^2 dx. \tag{2.96}$$

The integral in this expression can be integrated by parts as

$$\int_0^\ell \phi_i' \phi_i' dx = \phi_i' \phi_i \Big|_0^\ell - \int_0^\ell \phi_i(x) \phi_i''(x)\, dx. \tag{2.97}$$

By virtue of the boundary conditions at both ends, the first term is zero. Substitution of Eq. (2.40) into the last term (i.e., the integral) shows that

$$T \int_0^\ell \phi_i'^2 dx = m\omega_i^2 \int_0^\ell \phi_i^2 dx = M_i \omega_i^2, \tag{2.98}$$

---

[2] It is *not* true in general that the derivatives of mode shape functions form an orthogonal set.

where M_i was previously defined as the generalized mass and ω_i as the natural frequency of the ith mode. (It is noted that the ith generalized mass depends on the mode shape of the ith mode.) Thus, the potential energy becomes

$$P = \frac{1}{2} \sum_{i=1}^{\infty} M_i \omega_i^2 \xi_i^2. \tag{2.99}$$

The kinetic energy for a differential length of string is

$$dK = \frac{m}{2} \left[\left(\frac{\partial u}{\partial t} \right)^2 + \left(\frac{\partial v}{\partial t} \right)^2 \right] dx, \tag{2.100}$$

where the longitudinal displacement u was shown earlier to be less significant than the transverse displacement v and uncoupled from it. Thus,

$$\begin{aligned} dK &= \frac{m}{2} \left(\frac{\partial v}{\partial t} \right)^2 dx \\ &= \frac{m}{2} \left(\sum_{i=1}^{\infty} \phi_i \dot{\xi}_i \right)^2 dx. \end{aligned} \tag{2.101}$$

The total kinetic energy is obtained by integrating this over the entire string, and so

$$K = \frac{1}{2} \int_0^\ell m \left(\sum_{i=1}^{\infty} \phi_i \dot{\xi}_i \right)^2 dx. \tag{2.102}$$

With the double sum notation the kinetic energy simplifies to

$$\begin{aligned} K &= \frac{1}{2} \int_0^\ell \sum_{i=1}^{\infty} \sum_{j=1}^{\infty} \phi_i \dot{\xi}_i \phi_j \dot{\xi}_j m \, dx \\ &= \frac{1}{2} \sum_{i=1}^{\infty} \sum_{j=1}^{\infty} \dot{\xi}_i \dot{\xi}_j m \int_0^\ell \phi_i \phi_j dx. \end{aligned} \tag{2.103}$$

Because the mode shapes are orthogonal functions where

$$\int_0^\ell m \phi_i(x) \phi_j(x) \, dx = \begin{cases} 0 & (i \neq j), \\ M_i \neq 0 & (i = j), \end{cases} \tag{2.104}$$

the total kinetic energy becomes

$$K = \frac{1}{2} \sum_{i=1}^{\infty} M_i \dot{\xi}_i^2. \tag{2.105}$$

The "generalized equations of motion" can now be obtained by substitution of the kinetic energy of Eq. (2.105) and the potential energy of Eq. (2.99) into Lagrange's equations given as Eqs. (2.82). The resulting equations are then

$$M_i \left(\ddot{\xi}_i + \omega_i^2 \xi_i \right) = \Xi_i \qquad (i = 1, 2, \ldots). \tag{2.106}$$

2.1.7 *Generalized Force*

The generalized force, $\Xi_i(t)$, which appears on the right-hand side of the generalized equations of motion, represents the effective loading associated with all forces and moments not accounted for in P, which includes any nonconservative forces and moments. These forces and moments are most commonly identified as externally applied loads, which may or may not be a function of modal response. They will also include any dissipative loads such as those from dampers. To determine the contribution of distributed loads, denoted by $F(x, t)$, the virtual work will be computed as

$$\overline{\delta W} = \int_0^\ell F(x, t)\delta v(x, t)\, dx. \tag{2.107}$$

The term $\delta v(x, t)$ represents the virtual displacement, which can be written in terms of the generalized coordinates and mode shapes as

$$\delta v(x, t) = \sum_{i=1}^\infty \phi_i(x)\delta\xi_i(t), \tag{2.108}$$

where $\delta\xi_i(t)$ is an arbitrary increment in the ith generalized coordinate. Thus, the virtual work becomes

$$\begin{aligned}
\overline{\delta W} &= \int_0^\ell \sum_{i=1}^\infty F\phi_i\delta\xi_i\, dx \\
&= \sum_{i=1}^\infty \delta\xi_i \int_0^\ell F\phi_i\, dx.
\end{aligned} \tag{2.109}$$

Identifying the generalized force as

$$\Xi_i(t) = \int_0^\ell F(x, t)\phi_i(x)\, dx \tag{2.110}$$

one finds that the virtual work reduces to

$$\overline{\delta W} = \sum_{i=1}^\infty \Xi_i\, \delta\xi_i. \tag{2.111}$$

The loading $F(x, t)$ in the above development is a distributed load with units of force per unit length. If instead this loading is concentrated at one or more points, say as $F_c(t)$ with units of force acting at $x = x_c$ as shown in Fig. 2.8, then its functional representation must include the Dirac delta function, $\delta(x - x_c)$, which is similar to the impulse function in the time domain. In this case the distributed load can be written as

$$F(x, t) = F_c(t)\delta(x - x_c). \tag{2.112}$$

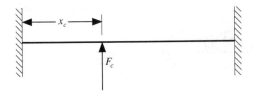

Figure 2.8 Concentrated force acting on string.

It may be recalled that the Dirac delta function can be thought of as the limiting case of a rectangular shape with area held constant and equal to unity as its width goes to zero. Thus, it may be defined by its integral property; for example, for $a < x_0 < b$

$$\int_a^b \delta(x - x_0)\,dx = 1,$$

$$\int_a^b f(x)\delta(x - x_0)\,dx = f(x_0). \tag{2.113}$$

As a consequence, the above integral expression for the generalized force can be applied to the concentrated load so that

$$\begin{aligned}
\Xi_i(t) &= \int_0^\ell F_c(t)\delta(x - x_c)\phi_i(x)\,dx \\
&= F_c(t)\phi_i(x_c).
\end{aligned} \tag{2.114}$$

Example 3: Calculation of Forced Response

An example of a dynamically loaded string will be considered to illustrate the generalized force computation and subsequent solution for the string displacement. The specific example will be a uniformly distributed load of simple harmonic amplitude (in time) shown in Fig. 2.9 with

$$F(x, t) = \overline{F}\sin(\omega t). \tag{2.115}$$

The initial string displacement and velocity will be taken as zero. Computation of the generalized force is simply

$$\begin{aligned}
\Xi_i &= \int_0^\ell \overline{F}\sin(\omega t)\sin\left(\frac{i\pi x}{\ell}\right)dx \\
&= \frac{\overline{F}\ell}{i\pi}\sin(\omega t)[1 - \cos(i\pi)].
\end{aligned} \tag{2.116}$$

Considering the even- and odd-indexed modes separately one has

$$\Xi_i = \begin{cases} \dfrac{2\overline{F}\ell}{i\pi}\sin(\omega t) & (i\text{ odd}), \\ 0 & (i\text{ even}). \end{cases} \tag{2.117}$$

With the above, the generalized equations of motion become

$$M_i\left(\ddot{\xi}_i + \omega_i^2\xi_i\right) = \begin{cases} \dfrac{2\overline{F}\ell}{i\pi}\sin(\omega t) & (i\text{ odd}), \\ 0 & (i\text{ even}). \end{cases} \tag{2.118}$$

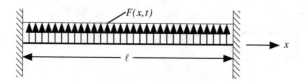

Figure 2.9   Distributed force $F(x, t)$ acting on string.

Since the initial conditions are

$$v(x, 0) = \frac{\partial v}{\partial t}(x, 0) = 0 \tag{2.119}$$

the even-indexed modes will not be excited because their generalized forces are also zero. For the odd-indexed modes the general solution to their equation of motion is

$$\xi_i = A_i \sin(\omega_i t) + B_i \cos(\omega_i t) + C_i \sin(\omega t). \tag{2.120}$$

It may be noted that the first two terms correspond to the homogeneous portion of the solution, whereas the third term represents the particular solution. In this example, the particular solution has the same form of time dependence as the generalized force.

To evaluate the constants A_i and B_i of the homogeneous solution, a procedure can be followed that is quite similar to the one used in Section 2.1.3 for solution of the homogeneous initial condition problem. The initial displacement of the present example can be written as

$$v(x, 0) = \sum_{i=1,3,\dots}^{\infty} \phi_i(x)\xi_i(0) = \sum_{i=1,3,\dots}^{\infty} B_i \sin\left(\frac{i\pi x}{\ell}\right) = 0. \tag{2.121}$$

Multiplying both sides of this relation by $\sin(j\pi x/\ell)$ and integrating over x from 0 to ℓ yields

$$\sum_{i=1,3,\dots}^{\infty} B_i \int_0^{\ell} \sin\left(\frac{i\pi x}{\ell}\right) \sin\left(\frac{j\pi x}{\ell}\right) dx = 0. \tag{2.122}$$

Applying the orthogonality property of the sine functions in the integrand indicates that

$$B_i = 0 \qquad (i \text{ odd}). \tag{2.123}$$

The same procedure can be applied to the initial velocity where

$$\frac{\partial v}{\partial t}(x, 0) = \sum_{i=1,3,\dots}^{\infty} \phi_i(x)\dot{\xi}_i(0) = \sum_{i=1,3,\dots}^{\infty} (A_i\omega_i + C_i\omega) \sin\left(\frac{i\pi x}{\ell}\right) = 0. \tag{2.124}$$

Again this relation can be multiplied by $\sin(j\pi x/\ell)$ and integrated over the string length. The orthogonality property in this case yields

$$A_i = -\frac{\omega C_i}{\omega_i} \qquad (i \text{ odd}). \tag{2.125}$$

The zero initial conditions thus require that the generalized coordinates of the odd-indexed modes be written as

$$\xi_i = C_i \left[\sin(\omega t) - \frac{\omega}{\omega_i} \sin(\omega_i t) \right] \qquad (i \text{ odd}). \tag{2.126}$$

The constants C_i of the particular integral can be determined by substitution of the generalized coordinate back into the generalized equations of motion. This yields

$$M_i C_i \left(\omega_i^2 - \omega^2\right) \sin(\omega t) = \frac{2\overline{F}\ell}{i\pi} \sin(\omega t). \tag{2.127}$$

With Eq. (2.48), $M_i = m\ell/2$ and the third constant becomes

$$C_i = \frac{4\overline{F}}{i\pi m \left(\omega_i^2 - \omega^2\right)}. \tag{2.128}$$

Thus, the string displacement can be written as a sum of the contributions from the odd-indexed modes. Recall that neither the excitation loading nor the initial conditions excite the even-indexed modes. Thus,

$$
\begin{aligned}
v(x,t) &= \sum_{i=1,3,\ldots}^{\infty} \xi_i(t)\phi_i(x) \\
&= \frac{4\overline{F}}{m\pi} \sum_{i=1,3,\ldots}^{\infty} \left[\frac{\sin(\omega t) - \frac{\omega}{\omega_i}\sin(\omega_i t)}{i\left(\omega_i^2 - \omega^2\right)} \right] \sin\left(\frac{i\pi x}{\ell}\right).
\end{aligned}
\tag{2.129}
$$

When the forcing frequency coincides with one of the natural frequencies, an interesting situation results. Take a typical term in the series solution of Eq. (2.129), and consider only its time-dependent part, for example,

$$
\frac{\sin(\omega t) - \frac{\omega}{\omega_i}\sin(\omega_i t)}{i\left(\omega_i^2 - \omega^2\right)}.
\tag{2.130}
$$

When $\omega \to \omega_i$, the term becomes indeterminate. To see what its value in the limit is, we let $\omega_i = \omega + \epsilon_i$, which gives

$$
\frac{\sin(\omega t) - \frac{\omega}{\omega + \epsilon_i}\sin[(\omega + \epsilon_i)t]}{i[(\omega + \epsilon_i)^2 - \omega^2]}.
\tag{2.131}
$$

Invoking l'Hopital's rule to take the limit as $\epsilon_i \to 0$, one obtains

$$
\frac{\sin(\omega t) - \omega t \cos(\omega t)}{2i\omega^2}.
\tag{2.132}
$$

The second term tends to infinity as time increases, with a linearly increasing amplitude. This phenomenon is called resonance and, because of its destructive nature, should be avoided. That is, when one excites a structure using harmonic excitation, the forcing frequency must not be too near any of the structure's natural frequencies.

Example 4: Calculation of Forced Response with Nonzero Initial Conditions

A second example will be considered to illustrate the treatment of a concentrated force and finite initial conditions. In this case a concentrated step-function force of magnitude F_0 will be applied to the center of the string as illustrated in Fig. 2.10. Recall that the unit step function, $1(t)$, is defined by

$$
1(t) = \begin{cases} 0 & (t < 0), \\ 1 & (t \geq 0). \end{cases}
\tag{2.133}
$$

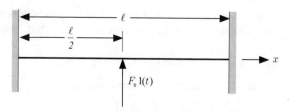

Figure 2.10   String with concentrated force at mid-span.

The initial shape of the string will be given as

$$v(x,0) = h \sin\left(\frac{4\pi x}{\ell}\right) \tag{2.134}$$

and the initial velocity as zero.

The generalized force can be determined from the integral of a distributed loading as

$$
\begin{aligned}
\Xi_i &= \int_0^\ell F(x,t)\phi_i(x)\,dx \\
&= \int_0^\ell F_0 1(t)\delta\left(x - \frac{\ell}{2}\right)\phi_i(x)\,dx \\
&= F_0 1(t)\phi_i\left(\frac{\ell}{2}\right) \\
&= F_0 1(t)\sin\left(\frac{i\pi}{2}\right).
\end{aligned}
\tag{2.135}
$$

Since

$$\sin\left(\frac{i\pi}{2}\right) = \begin{cases} 0 & (i \text{ even}), \\ (-1)^{\frac{i-1}{2}} & (i \text{ odd}), \end{cases} \tag{2.136}$$

the generalized equations of motion become

$$M_i\left(\ddot{\xi}_i + \omega_i^2\xi_i\right) = \begin{cases} 0 & (i \text{ even}), \\ F_0 1(t)(-1)^{\frac{i-1}{2}} & (i \text{ odd}). \end{cases} \tag{2.137}$$

The corresponding general solutions are

$$
\begin{aligned}
\xi_i &= A_i \sin(\omega_i t) + B_i \cos(\omega_i t) & (i \text{ even}), \\
\xi_i &= A_i \sin(\omega_i t) + B_i \cos(\omega_i t) + C_i & (i \text{ odd}).
\end{aligned}
\tag{2.138}
$$

Consider the finite initial displacement

$$
\begin{aligned}
v(x,0) &= \sum_{i=1}^\infty \xi_i(0)\phi_i(x) \\
&= \sum_{i=2,4,\dots}^\infty B_i \sin\left(\frac{i\pi x}{\ell}\right) + \sum_{i=1,3,\dots}^\infty (B_i + C_i)\sin\left(\frac{i\pi x}{\ell}\right) \\
&= h\sin\left(\frac{4\pi x}{\ell}\right).
\end{aligned}
\tag{2.139}
$$

This last equality will be multiplied by $\sin(j\pi x/\ell)$ and integrated over the length of the string to yield

$$
\begin{aligned}
h\int_0^\ell \sin\left(\frac{4\pi x}{\ell}\right)\sin\left(\frac{j\pi x}{\ell}\right)dx &= \sum_{i=2,4,\dots}^\infty B_i \int_0^\ell \sin\left(\frac{i\pi x}{\ell}\right)\sin\left(\frac{j\pi x}{\ell}\right)dx \\
&\quad + \sum_{i=1,3,\dots}^\infty (B_i + C_i)\int_0^\ell \sin\left(\frac{i\pi x}{\ell}\right) \\
&\qquad \times \sin\left(\frac{j\pi x}{\ell}\right)dx.
\end{aligned}
\tag{2.140}
$$

These integrals can easily be evaluated by noting the orthogonality property of the sine functions. The result gives the following values for the constants B_i:

$$B_4 = h,$$

$$B_i = \begin{cases} 0 & (i \text{ even but } \neq 4), \\ -C_i & (i \text{ odd}). \end{cases} \tag{2.141}$$

The zero initial velocity requires that

$$\frac{\partial v}{\partial t}(x, 0) = \sum_{i=1}^{\infty} \dot{\xi}_i(0)\phi_i(x) = \sum_{i=1}^{\infty} \omega_i A_i \sin\left(\frac{i\pi x}{\ell}\right) = 0. \tag{2.142}$$

Multiplication by $\sin(j\pi x/\ell)$ and integration will result in determining that $A_i = 0$ for all i. These results can be summarized by noting that $\xi_i = 0$ for all even values of i except

$$\xi_4 = h\cos(\omega_4 t), \tag{2.143}$$

and for odd i

$$\xi_i = C_i\left[1 - \cos(\omega_i t)\right] \qquad (i \text{ odd}). \tag{2.144}$$

The C_is can be determined by substitution of the odd generalized coordinates back into the equations of motion,

$$M_i C_i \omega_i^2 = F_0(-1)^{\frac{i-1}{2}} \qquad (t \geq 0). \tag{2.145}$$

Given that $M_i = m\ell/2$, this yields

$$C_i = \frac{2\ell F_0(-1)^{\frac{i-1}{2}}}{T(i\pi)^2}, \tag{2.146}$$

so that the complete string displacement becomes

$$\begin{aligned} v(x, t) &= \sum_{i=1}^{\infty} \xi_i(t)\phi_i(x) \\ &= h\cos(\omega_4 t)\sin\left(\frac{4\pi x}{\ell}\right) \\ &\quad + \frac{2\ell F_0}{T\pi^2} \sum_{i=1,3,\ldots}^{\infty} \frac{(-1)^{\frac{i-1}{2}}}{i^2}\left[1 - \cos(\omega_i t)\right]\sin\left(\frac{i\pi x}{\ell}\right). \end{aligned} \tag{2.147}$$

The first term is thus the response due to the initial displacement, and the sum over the odd-indexed modes is the response due to the forcing function.

2.2   Uniform Beam Torsional Dynamics

Now that the fundamental aspects of structural dynamics analysis have been considered for the uniform string problem, these concepts will be applied to the dynamics of beam torsional deformation. The beam has many more of the characteristics of typical aeronautical structures. Indeed, high-aspect-ratio wings and helicopter rotor blades are frequently idealized as beams, especially in preliminary design. Even for low-aspect-ratio wings, although a plate model is more realistic, the bending and torsional deformation can be approximated by use of beam theory with adjusted stiffness coefficients.

In an effort to retain a level of simplicity that promotes tractability, the torsional rigidity of St. Venant theory, denoted GJ, will be taken as given. For homogeneous and isotropic beams, G denotes the shear modulus and J is a constant that depends only on the geometry of the cross section. For such beams J can be determined by solving a boundary-value problem over the cross-sectional area, which requires finding the cross-sectional warping caused by torsion. Although analytical solutions for this problem are available for some simple cross-sectional geometries, solving for the cross-sectional warping and torsional stiffness is not a trivial exercise in general. For nonhomogeneous, anisotropic beams one may also use the symbol GJ to denote an effective torsional rigidity, which can be determined by solving a far more involved boundary-value problem over the cross-sectional area.

2.2.1   *Equation of Motion*

The beam will be considered initially to have nonuniform properties along the x-axis, which is chosen to coincide with the elastic axis for the beam. In our idealized model for beams, this axis is assumed to be straight and, by definition, corresponds to the locus of the cross-sectional shear centers. This choice of x-axis structurally uncouples torsion and transverse displacements caused by bending for isotropic beams. For composite beams, this choice of x-axis uncouples torsion and transverse shear deformation, but torsion and transverse displacements may remain coupled depending on whether the beam has bending–torsion elastic coupling. For a further simplification, we assume that cross-sectional mass centroids lie along the elastic axis, in which case transverse motions due to bending are inertially uncoupled from torsional motion. The elastic twisting deflection, θ, will be positive in a right-handed sense about this axis as illustrated in Fig. 2.11. In contrast, the twisting moment, denoted by T, is the structural torque, that is, the resultant moment of the tractions on a cross-sectional face about the elastic axis. Recall that an outward-directed normal on the positive x face is directed to the right, whereas an outward-directed normal on the negative x face is directed to the left. Thus, the direction of a positive torque tends to rotate the positive x face in a direction that is positive in the right-hand sense and the negative x face in a direction that is negative in the right-hand sense, as depicted in Fig. 2.11. This will affect the boundary conditions, as noted below.

Letting $\rho I_p dx$ be the polar mass moment of inertia about the x-axis of the differential beam segment in Fig. 2.12, one can obtain the equation of motion by equating the resultant twisting moment on both faces of the segment to the rate of change of the segment's angular momentum about the elastic axis. This yields

$$T + \frac{\partial T}{\partial x}dx - T = \rho I_p dx \frac{\partial^2 \theta}{\partial t^2}, \tag{2.148}$$

or

$$\frac{\partial T}{\partial x} = \rho I_p \frac{\partial^2 \theta}{\partial t^2}, \tag{2.149}$$

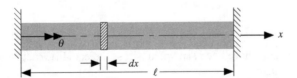

Figure 2.11   Beam undergoing torsional deformation.

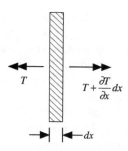

Figure 2.12   Cross-sectional slice of beam undergoing torsional deformation.

where

$$\rho I_p = \int_{\mathcal{A}} \rho(y^2 + z^2)\, d\mathcal{A}. \tag{2.150}$$

Here $\mathcal{A}$ is the cross section of the beam, y and z are cross-sectional Cartesian coordinates, and ρ is the mass density of the beam. When ρ is constant over the cross section, then I_p is the area polar moment of inertia per unit length. When ρ varies over the cross section, one may interpret ρI_p as the mass polar moment of inertia for the cross section.

The twisting moment can be written in terms of the twist rate and the Saint-Venant torsional rigidity as

$$T = GJ \frac{\partial \theta}{\partial x}. \tag{2.151}$$

Substituting these expressions into Eq. (2.149), one obtains the partial differential equation of motion for the nonuniform beam given by

$$\frac{\partial}{\partial x}\left(GJ\frac{\partial \theta}{\partial x}\right) = \rho I_p \frac{\partial^2 \theta}{\partial t^2}. \tag{2.152}$$

For the special case of uniform beams this relation simplifies to the one-dimensional wave equation

$$\frac{GJ}{\rho I_p}\frac{\partial^2 \theta}{\partial x^2} = \frac{\partial^2 \theta}{\partial t^2}. \tag{2.153}$$

Other than the constants that multiply the second partial derivatives, this is the same equation that governed the dynamic behavior of the string. Thus, all the previously discussed properties of standing and traveling waves will exist here as well. However, as will be discussed in detail below, there are more interesting possibilities for the boundary conditions.

To establish these properties the separation of variables method will be applied as

$$\theta(x, t) = X(x)Y(t), \tag{2.154}$$

which, when substituted into the wave equation, yields

$$\frac{X''}{X} = \frac{\rho I_p}{GJ}\frac{\ddot{Y}}{Y}. \tag{2.155}$$

Since the dependencies on x and t have been separated across the equality, each side must equal a constant, say $-\alpha^2$, so that

$$\frac{X''}{X} = \frac{\rho I_p}{GJ}\frac{\ddot{Y}}{Y} = -\alpha^2. \tag{2.156}$$

Figure 2.13   Clamped end of a beam.

Two ordinary differential equations then follow from this, namely,

$$X''(x) + \alpha^2 X(x) = 0,$$

$$\ddot{Y}(t) + \alpha^2 \frac{GJ}{\rho I_p} Y(t) = 0.$$

(2.157)

Note the similarity with Eqs. (2.17). For $\alpha \neq 0$, Eqs. (2.157) have solutions that can be written as

$$X(x) = A \sin(\alpha x) + B \cos(\alpha x),$$

$$Y(t) = C \sin\left(\sqrt{\frac{GJ}{\rho I_p}} \alpha t\right) + D \cos\left(\sqrt{\frac{GJ}{\rho I_p}} \alpha t\right).$$

(2.158)

To complete the solution the constants A and B can be determined from the boundary conditions at the ends of the beam, and C and D can be found as a function of the initial beam deflection and rate of deflection. The special case of $\alpha = 0$ is very important and will be addressed in more detail in Section 2.2.3.

2.2.2   Boundary Conditions

There are four different boundary conditions that can be imposed at the ends of the beam for determination of the constants A and B. For any given beam only one boundary condition is required at each end.

Clamped End
In this case (see Fig. 2.13) the end of the beam is assumed to be "built-in" or "cantilevered" at a wall. As a consequence there will be no rotation due to elastic twist at the beam end, and the boundary condition is

$$\theta(\ell, t) = 0 = X(\ell) Y(t),$$

(2.159)

which is identically satisfied when

$$X(\ell) = 0.$$

(2.160)

Free End
When the end of the beam is free (see Fig. 2.14) there are no external loads acting on the beam. Therefore the twisting moment at the end must be zero:

$$T(\ell, t) = GJ \frac{\partial \theta}{\partial x}(\ell, t) = 0,$$

(2.161)

Figure 2.14   Free end of a beam.

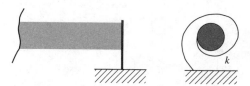

Figure 2.15 Elastically restrained end of a beam.

or, for uniform beams

$$\frac{\partial \theta}{\partial x}(\ell, t) = X'(\ell)Y(t) = 0. \tag{2.162}$$

Thus, the specific condition to be satisfied is

$$X'(\ell) = 0. \tag{2.163}$$

Elastic Constraint

An elastic constraint (see Fig. 2.15) can be treated as a linearly elastic torsional spring attached to the end of the beam. The twisting moment at the beam end must be equal and opposite to the spring reaction for any finite deflection of the end so that

$$
\begin{aligned}
T(\ell, t) &= GJ\frac{\partial \theta}{\partial x}(\ell, t) \\
&= -k\theta(\ell, t),
\end{aligned} \tag{2.164}
$$

where k is the spring constant. Thus,

$$GJX'(\ell)Y(t) = -kX(\ell)Y(t), \tag{2.165}$$

which requires that

$$GJX'(\ell) = -kX(\ell). \tag{2.166}$$

The reader should verify that the same type of boundary condition at the other end would yield

$$GJX'(0) = kX(0), \tag{2.167}$$

where the sign change comes about by virtue of the switch in direction noted above for a positive twisting moment.

Inertial Constraint

Here we consider a beam with a rigid body attached to its right end (see Fig. 2.16). This rigid body has a mass moment of inertia about the beam elastic axis denoted by I_c, which contributes a concentrated rotational inertia effect. The twisting moment at the beam end must be equal and opposite to the inertial reaction of the concentrated inertia for any

$$x = \ell$$

Figure 2.16 Inertially restrained end of a beam.

finite angular acceleration of the end. Therefore,

$$T(\ell, t) = GJ\frac{\partial \theta}{\partial x}(\ell, t) = -I_c\frac{\partial^2 \theta}{\partial t^2}(\ell, t), \tag{2.168}$$

so that

$$GJX'(\ell)Y(t) = -I_cX(\ell)\ddot{Y}(t). \tag{2.169}$$

From the functional form of $Y(t)$ as established from the separation procedure, it can be noted that

$$\ddot{Y}(t) = -\alpha^2\frac{GJ}{\rho I_p}Y(t). \tag{2.170}$$

Substitution into the preceding condition yields

$$GJX'(\ell)Y(t) = \alpha^2\frac{GJ}{\rho I_p}I_cX(\ell)Y(t), \tag{2.171}$$

which requires that

$$\rho I_pX'(\ell) = \alpha^2 I_cX(\ell). \tag{2.172}$$

As above, the reader should verify that the same type of boundary condition at the other end would yield

$$\rho I_pX'(0) = -\alpha^2 I_cX(0). \tag{2.173}$$

2.2.3   *Example Solutions for Mode Shapes and Frequencies*

In this section we consider several examples of the calculation of natural frequencies and mode shapes of vibrating beams in torsion. We begin with the clamped–free case. Next, we consider the free–free case, illustrating the concept of the rigid-body mode. Finally, we consider a case that requires numerical solution of the transcendental characteristic equation, a beam clamped at its root and restrained with a rotational spring at its tip.

Example 5: Solution for Clamped–Free Beam

To illustrate the application of these boundary conditions, consider the case of a uniform beam that is clamped at $x = 0$ and free at $x = \ell$, as shown in Fig. 2.17. The boundary conditions for this case are

$$X(0) = X'(\ell) = 0. \tag{2.174}$$

Figure 2.17   Schematic of clamped–free beam undergoing torsion.

Recall that the general solution was previously determined as

$$\theta(x, t) = X(x)Y(t), \tag{2.175}$$

where X and Y are given in Eqs. (2.158). For $\alpha \neq 0$ the first of those equations has the solution

$$X(x) = A \sin(\alpha x) + B \cos(\alpha x). \tag{2.176}$$

It is apparent that the boundary conditions lead to the following:

$$X(0) = 0 \text{ requires } B = 0,$$
$$X'(\ell) = 0 \text{ requires } A\alpha \cos(\alpha \ell) = 0. \tag{2.177}$$

If $A = 0$ a trivial solution will be obtained, such that the deflection will be identically zero. Since $\alpha \neq 0$, a nontrivial solution requires that

$$\cos(\alpha \ell) = 0. \tag{2.178}$$

This is called the "characteristic equation," the solutions of which consist of a denumerably infinite set called the "eigenvalues" and are given by

$$\alpha_i \ell = \frac{(2i - 1)\pi}{2} \qquad (i = 1, 2, \ldots). \tag{2.179}$$

The $Y(t)$ portion of the general solution can be observed to have the form of simple harmonic motion, as indicated in Eq. (2.170), so that the natural frequency is

$$\omega = \alpha \sqrt{\frac{GJ}{\rho I_p}}. \tag{2.180}$$

Since α can have only specific values, the frequencies will also take on specific numerical values given by

$$\omega_i = \alpha_i \sqrt{\frac{GJ}{\rho I_p}} = \frac{(2i - 1)\pi}{2\ell} \sqrt{\frac{GJ}{\rho I_p}}. \tag{2.181}$$

These are the natural frequencies of the beam. Associated with each frequency is a "mode shape" as determined from the x-dependent portion of the general solution. The mode shapes (or eigenfunctions) can be written as

$$\phi_i(x) = \sin(\alpha_i x) = \sin\left[\frac{(2i - 1)\pi x}{2\ell}\right] \tag{2.182}$$

or any constant times $\phi_i(x)$. The first three of these mode shapes are plotted in Fig. 2.18. The zero derivative at the free end is indicative of the vanishing twisting moment at the free end.

Example 6: Solution for Free–Free Beam

A second example, which exhibits both elastic motion as described above and motion as a rigid body, is the case of a beam that is free at both ends as shown in Fig. 2.19. The boundary conditions are

$$X'(0) = X'(\ell) = 0. \tag{2.183}$$

From the general solution for $X(x)$ in Eqs. (2.158), one finds that for $\alpha \neq 0$

$$X'(x) = A\alpha \cos(\alpha x) - B\alpha \sin(\alpha x). \tag{2.184}$$

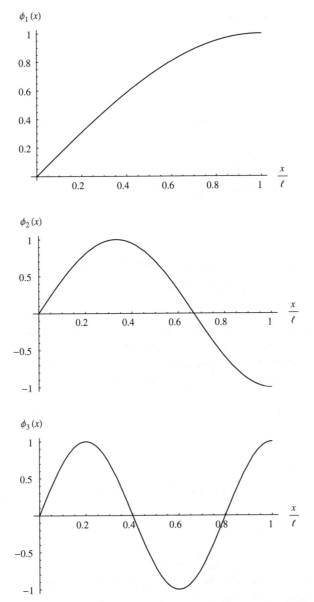

Figure 2.18   First three mode shapes for clamped–free beam vibrating in torsion.

Figure 2.19   Schematic of free–free beam undergoing torsion.

Thus, the condition at $x = 0$ requires that

$$A\alpha = 0. \tag{2.185}$$

For $A = 0$ the condition at $x = \ell$ requires that

$$\sin(\alpha\ell) = 0 \tag{2.186}$$

since a null solution ($\theta \equiv 0$) is obtained if $B = 0$. This characteristic equation is satisfied by

$$\alpha_i\ell = i\pi \qquad (i = 1, 2, \ldots), \tag{2.187}$$

and the corresponding natural frequencies become

$$\omega_i = \frac{i\pi}{\ell}\sqrt{\frac{GJ}{\rho I_p}}. \tag{2.188}$$

The associated mode shapes are determined from the corresponding $X(x)$ as

$$\phi_i(x) = \cos(\alpha_i x) = \cos\left(\frac{i\pi x}{\ell}\right). \tag{2.189}$$

These frequencies and mode shapes describe the normal mode of vibration for the elastic degrees of freedom of the free–free beam in torsion.

Now, if in the above analysis the separation constant, α, is taken as zero, then the governing ordinary differential equations are changed to

$$\frac{X''}{X} = \frac{\rho I_p}{GJ}\frac{\ddot{Y}}{Y} = 0 \tag{2.190}$$

or

$$X''(x) = 0 \text{ and } \ddot{Y}(t) = 0. \tag{2.191}$$

The general solutions to these equations can be written as

$$X(x) = ax + b,$$
$$Y(t) = ct + d. \tag{2.192}$$

The arbitrary constants, a and b, in the spatially dependent portion of the solution can again be determined from the boundary conditions. For the present case of the free–free beam the conditions are

$$X'(0) = 0 \text{ requires } a = 0,$$
$$X'(\ell) = 0 \text{ requires } a = 0. \tag{2.193}$$

Because both conditions are satisfied without imposing any restrictions on the constant b, this constant can be anything, which implies that the torsional deflection can be nontrivial for $\alpha = 0$. From $X(x)$ with $a = 0$ it is apparent that the corresponding value of θ will be independent of the coordinate x. This means that this motion for $\alpha = 0$ is a "rigid-body" rotation of the beam.

The time-dependent solution for this motion, $Y(t)$, is also different from that obtained for the elastic motion. Primarily it can be noted that the motion is not oscillatory; thus, the rigid-body natural frequency is zero. The arbitrary constants, c and d, can be obtained from the initial values of the rigid-body orientation and angular velocity. To summarize the complete solution for the free–free beam in torsion, a set of generalized coordinates can be defined by

$$\theta(x,t) = \sum_{i=0}^{\infty} \phi_i(x)\xi_i(t), \qquad (2.194)$$

where

$$\phi_0 = 1,$$

$$\phi_i = \cos\left(\frac{i\pi x}{\ell}\right) \qquad (i = 1, 2, \ldots). \qquad (2.195)$$

The first three elastic mode shapes are plotted in Fig. 2.20. The zero derivative at both ends is indicative of the vanishing twisting moment there. The natural

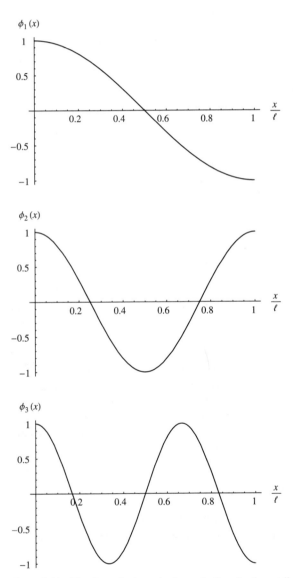

Figure 2.20   First three elastic mode shapes for free–free beam vibrating in torsion.

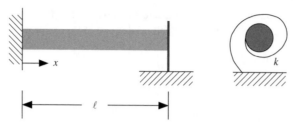

Figure 2.21   Schematic of torsion problem with spring.

frequencies associated with these mode shapes are

$$\omega_0 = 0,$$

$$\omega_i = \frac{i\pi}{\ell}\sqrt{\frac{GJ}{\rho I_p}} \qquad (i = 1, 2, \ldots). \tag{2.196}$$

It may be noted that the rigid-body generalized coordinate, $\xi_0(t)$, represents the radian measure of the rigid-body rotation of the beam about the x axis.

Note that a quick way to verify the existence of a rigid-body mode is to substitute $\omega = 0$ and $X =$ a constant into the differential equation and boundary conditions for X. A rigid-body mode exists if and only if all are satisfied.

Example 7: Solution for Clamped–Spring-Restrained Beam

A final example for beam torsion is given by the system in Fig. 2.21. The beam is clamped at the root ($x = 0$) end, and the other end is restrained with a rotational spring having spring constant $k = \zeta GJ/\ell$, where ζ is a dimensionless parameter. The boundary conditions on X are thus

$$X(0) = 0,$$

$$GJX'(\ell) = -kX(\ell) = -\frac{GJ}{\ell}\zeta X(\ell). \tag{2.197}$$

When these boundary conditions are substituted into the general solution found in Eqs. (2.158), one sees that the first condition requires that $B = 0$; the second condition, along with the requirement for a nontrivial solution, leads to

$$\zeta \tan(\alpha\ell) + \alpha\ell = 0. \tag{2.198}$$

This transcendental equation has a denumerably infinite set of roots that cannot be found in closed form. However, as many of these roots as desired can be found using numerical procedures found in commercially available software packages such as Mathematica, Maple, or MATLAB. These roots of Eq. (2.198) will be functions of ζ, and the first four such roots are plotted versus ζ in Fig. 2.22. Denoting these roots by α_i, with $i = 1, 2, \ldots$, one obtains the corresponding natural frequencies

$$\omega_i = \alpha_i\sqrt{\frac{GJ}{\rho I_p}} \qquad (i = 1, 2, \ldots). \tag{2.199}$$

From the plots (and from Eq. 2.198) we note that as ζ tends toward zero, α_1 tends toward $\pi/2$, which means that the fundamental natural frequency is

$$\omega_1 = \frac{\pi}{2\ell}\sqrt{\frac{GJ}{\rho I_p}} \qquad (\zeta \to 0). \tag{2.200}$$

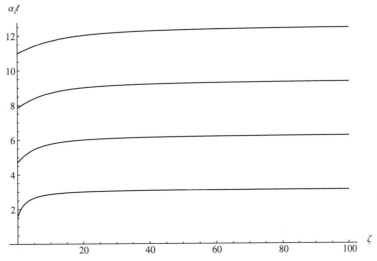

Figure 2.22 Plot of the lowest values of $\alpha_i \ell$ versus ζ for clamped–spring-restrained beam in torsion.

which is the natural frequency of a clamped–free beam in torsion (as shown above). One can also show that as ζ tends to infinity, α_1 tends toward π so that the fundamental natural frequency is

$$\omega_1 = \frac{\pi}{\ell} \sqrt{\frac{GJ}{\rho I_p}} \qquad (\zeta \to \infty), \tag{2.201}$$

which is the natural frequency of a clamped–clamped beam in torsion. The determination of the natural frequencies of a clamped–clamped beam in torsion is left as an exercise for the reader.

To obtain the corresponding mode shapes, one takes the solutions for α_i and substitutes back into X, recalling that we can arbitrarily set $A = 1$ and that $B = 0$. The resulting mode shape is

$$\phi_i = \sin(\alpha_i x) \qquad (i = 1, 2, \ldots). \tag{2.202}$$

The first three modes for $\zeta = 1$ are shown in Fig. 2.23 and 2.24. As expected, neither the twist angle nor its derivative are equal to zero at the tip.

2.3 Uniform Beam Bending Dynamics

The free vibration of a beam in transverse bending motion is often referred to as transverse vibration. This type of motion differs from the transverse string dynamics and beam torsional dynamics in that the governing equations of motion are of a different mathematical form. Although these equations are different, their solutions are obtained in a similar manner and exhibit similar physical characteristics. It should also be observed that, whereas most aerospace structures will experience combined or simultaneous bending and torsional dynamic behavior, we have here chosen certain configuration variables to uncouple these types of motion.

2.3.1 Equation of Motion

As in the case of torsion the beam will initially be treated as having nonuniform properties along the x axis. The x axis will be taken as the line of the individual

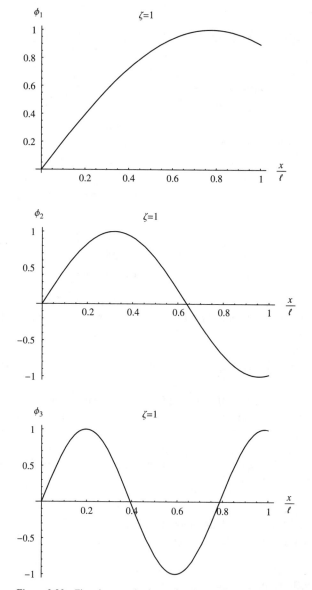

Figure 2.23   First three mode shapes for clamped–spring-restrained beam in torsion, $\zeta = 1$.

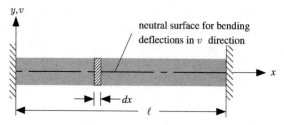

Figure 2.24   Schematic of beam for bending dynamics.

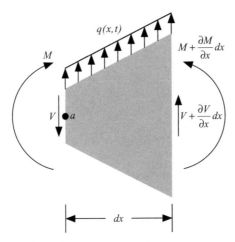

Figure 2.25   Schematic of differential beam segment.

cross-sectional neutral axes associated with pure bending in and normal to the plane of the diagram in Fig. 2.24. For simplicity, however, we will only consider uncoupled bending in the x–y plane, thus excluding initially twisted beams from the development. The bending deflections are denoted by $v(x, t)$ in the y direction. The x axis is presumed to be straight, thus excluding initially curved beams. We will continue to assume for now that the properties of the beam allow the x axis to be chosen so that bending and torsion are both structurally and inertially uncoupled. Finally, the transverse beam displacement, v, will be presumed small to permit a linearly elastic representation of the deformation.

A free-body diagram for the differential beam segment shown in Fig. 2.25 includes the shear force, V, and the bending moment, M. Recall from our earlier discussion on torsion that an outward-directed normal on the positive x face is directed to the right, and an outward-directed normal is directed to the left on the negative x face. By this convention, V is the resultant of the transverse shear stresses in the positive y direction (upward in Fig. 2.25) on a positive x cross-section face and in the negative y direction on a negative x cross-section face. In other words, a positive shear force tends to displace the positive x face upward and the negative x face downward, as depicted in Fig. 2.25. The bending moment, M, is the moment of the longitudinal stresses about a line parallel to the z-axis (out of the paper in Fig. 2.25) at the intersection between the cross-sectional plane and the neutral surface. Thus, a positive bending moment tends to rotate the positive x face positively about the z-axis (in the right-handed sense) and the negative x face negatively about the z-axis. This will affect the boundary conditions, as noted below. The distributed loading (with units of force per unit length) is denoted by q. The equation of motion for transverse beam displacements can be obtained by setting the resultant force on the segment equal to its mass times its acceleration, which yields

$$q(x, t)\, dx - V + \left(V + \frac{\partial V}{\partial x} dx \right) = m\, dx \frac{\partial^2 v}{\partial t^2} \tag{2.203}$$

and leads to

$$-\frac{\partial V}{\partial x} + m \frac{\partial^2 v}{\partial t^2} = q(x, t), \tag{2.204}$$

where $m = \rho A$ is the mass per unit length, ρ is the material density, and A is the cross-sectional area. We must also consider the moment equation. We note here that the

cross-sectional rotational inertia (about an axis normal to the page) will be ignored because it has a small effect. Taking a counterclockwise moment as positive, we sum the
moments about the point a to obtain

$$-M + \left(M + \frac{\partial M}{\partial x}dx\right) + \left(V + \frac{\partial V}{\partial x}dx\right)dx - \frac{1}{2}m\frac{\partial^2 v}{\partial t^2}dx^2 = 0, \tag{2.205}$$

which, after we neglect the higher-order differentials (i.e., higher powers of dx), becomes

$$\frac{\partial M}{\partial x} + V = 0. \tag{2.206}$$

Recall that the bending moment is proportional to the local curvature and so

$$M = EI\frac{\partial^2 v}{\partial x^2}, \tag{2.207}$$

where EI may be regarded as the effective bending stiffness of the beam at a particular cross
section. For homogeneous and isotropic beams, E is Young's modulus and I is the cross-
sectional area moment of inertia about the z axis for a particular cross section. Substitution
of Eq. (2.207) into Eq. (2.206) and of the resulting equation into Eq. (2.204) yields the
partial differential equation of motion for a nonuniform beam as

$$\frac{\partial^2}{\partial x^2}\left(EI\frac{\partial^2 v}{\partial x^2}\right) + m\frac{\partial^2 v}{\partial t^2} = q(x,t). \tag{2.208}$$

In the following sections, we will be treating the special case of free vibration for which
$q(x,t) = 0$. Also, for simplicity we will specialize the equations for the case of spanwise
uniformity of all properties, for which the parameter a can be defined as a constant such
that

$$a^4 = \frac{EI}{m}, \tag{2.209}$$

to simplify the equation of motion to

$$a^4\frac{\partial^4 v}{\partial x^4} + \frac{\partial^2 v}{\partial t^2} = 0. \tag{2.210}$$

2.3.2 *General Solutions*

A solution to the equation of motion for transverse beam vibrations can be obtained
by a separation of the independent variables. This separation will be denoted as

$$v(x,t) = X(x)Y(t), \tag{2.211}$$

which when substituted into the equation of motion yields

$$\frac{X''''}{X} = -\frac{1}{a^4}\frac{\ddot{Y}}{Y}. \tag{2.212}$$

Since the dependencies on x and t have been separated across the equality, each side must
equal a constant, say α^4. The resulting ordinary differential equations then become

$$X'''' - \alpha^4 X = 0,$$
$$\ddot{Y} + a^4\alpha^4 Y = 0. \tag{2.213}$$

For $\alpha \neq 0$, the general solution to the second (the time-dependent) equation can be written as in the cases for the string and beam torsion, namely

$$Y(t) = A \sin(\omega t) + B \cos(\omega t), \tag{2.214}$$

where it is clear from the second of Eqs. (2.213) that

$$\omega = \alpha^2 a^2 = \alpha^2 \sqrt{\frac{EI}{m}} = (\alpha \ell)^2 \sqrt{\frac{EI}{m\ell^4}}. \tag{2.215}$$

For $\alpha \neq 0$, the general solution to the spatially dependent equation can be obtained by presuming a solution of the form

$$X(x) = \exp(\lambda x). \quad \left(e^{\lambda x} \right) \tag{2.216}$$

Substitution of this assumed form into the fourth-order differential equation for $X(x)$ yields

$$\lambda^4 - \alpha^4 = 0, \tag{2.217}$$

which can be factored to

$$(\lambda - i\alpha)(\lambda + i\alpha)(\lambda - \alpha)(\lambda + \alpha), \tag{2.218}$$

which indicates a general solution of the form

$$X(x) = C_1 \exp(i\alpha x) + C_2 \exp(-i\alpha x) + C_3 \exp(\alpha x) + C_4 \exp(-\alpha x). \tag{2.219}$$

Rewriting the exponential functions as trigonometric and hyperbolic sine and cosine functions yields an alternative form of the general solution as

$$X(x) = D_1 \sin(\alpha x) + D_2 \cos(\alpha x) + D_3 \sinh(\alpha x) + D_4 \cosh(\alpha x). \tag{2.220}$$

Eventual determination of the constants D_i ($i = 1, 2, 3,$ and 4) and α will require specification of appropriate boundary conditions. To facilitate this procedure this last solution form can be rearranged to provide in some cases a slight advantage in the algebra, so that

$$\begin{aligned} X(x) = {} & E_1[\sin(\alpha x) + \sinh(\alpha x)] + E_2[\sin(\alpha x) - \sinh(\alpha x)] \\ & + E_3[\cos(\alpha x) + \cosh(\alpha x)] + E_4[\cos(\alpha x) - \cosh(\alpha x)]. \end{aligned} \tag{2.221}$$

To complete the solution the constants A and B can be determined from the initial deflection and rate of deflection of the beam. The remaining constants, C_i, D_i, or E_i ($i = 1, 2, 3,$ and 4), can be evaluated from the boundary conditions, which must be imposed at each end of the beam. As was true for torsion, the very important special case of $\alpha = 0$ is connected with rigid-body modes for beam bending and is addressed in more detail in Section 2.3.4.

2.3.3 *Boundary Conditions*

For the beam bending problem it is necessary to impose two boundary conditions at each end of the beam. A boundary condition can be any linear, homogeneous relation involving the beam deflection and one or more of its partial derivatives. Although it is not a mathematical requirement, the particular combination of conditions to be specified at a beam end should represent a physically realizable constraint. The various derivatives of the beam deflection can be associated with particular beam states at any arbitrary point along

$x = \ell$

Figure 2.26   Schematic of pinned end condition.

the beam. There are four such states of practical interest:

1. deflection $= v(x, t) = X(x)Y(t)$,
2. slope $= \frac{\partial v}{\partial x}(x, t) = X'(x)Y(t)$,
3. bending moment $= M(x, t) = EI\frac{\partial^2 v}{\partial x^2}(x, t) = EIX''(x)Y(t)$,
4. shear $= V(x, t) = -EI\frac{\partial^3 v}{\partial x^3}(x, t) = -EIX'''(x)Y(t)$.

It should be noted when relating these beam states that the positive convention for deflection and slope is the same at both ends of the beam. In contrast, the shear and bending moment sign conventions differ at opposite beam ends as illustrated by the free-body differential beam element used to obtain the equation of motion shown in Fig. 2.25.

The most common conditions that can occur at the beam ends involve vanishing pairs of individual states. Typical of such conditions are the following classical configurations:

- Simply-supported, hinged or pinned end, which indicates zero deflection and bending moment, is denoted by the triangular symbol in Fig. 2.26 and has $v(\ell, t) = M(\ell, t) = 0$ so that $X(\ell) = X''(\ell) = 0$.
- Cantilever or built-in end, which implies zero deflection and slope, is illustrated in Fig. 2.13 and has $v(\ell, t) = \frac{\partial v}{\partial x}(\ell, t) = 0$ so that $X(\ell) = X'(\ell) = 0$.
- Free end, which corresponds to zero bending moment and shear, is illustrated in Fig. 2.14 and has $M(\ell, t) = V(\ell, t) = 0$ so that $X''(\ell) = X'''(\ell) = 0$.
- Sliding end, which corresponds to zero slope and shear, is illustrated in Fig. 2.27 and has $\frac{\partial v}{\partial x}(\ell, t) = V(\ell, t) = 0$ so that $X'(\ell) = X'''(\ell) = 0$.

All of these conditions can occur in the same form at $x = 0$.

In addition to these zero-state conditions, the boundary conditions can correspond to linear constraint reactions associated with elastic and inertial elements. These types of conditions were previously observed for the torsional dynamics of beams. They may occur alone or in conjunction with others. The boundary conditions associated with these constraint reactions are of four basic types:

1. translational elastic constraint,
2. rotational elastic constraint,
3. translational inertia constraint, and
4. rotational inertia constraint.

$x = \ell$

Figure 2.27   Schematic of sliding end condition.

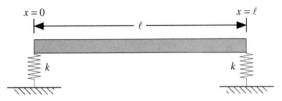

Figure 2.28   Schematic of translational spring end conditions.

Translational Elastic Constraint

A translational elastic constraint corresponds to a spring reaction force that is equated to the shear force, as shown in Fig. 2.28. At the left end

$$EI\frac{\partial^3 v}{\partial x^3}(0, t) = -kv(0, t) \qquad \rightarrow \qquad EIX'''(0) = -kX(0). \tag{2.222}$$

At the right end

$$EI\frac{\partial^3 v}{\partial x^3}(\ell, t) = kv(\ell, t) \qquad \rightarrow \qquad EIX'''(\ell) = kX(\ell). \tag{2.223}$$

As previously observed, it is seen here that the constraint relations differ in sign at opposite beam ends. It should also be noted that the above conditions must be augmented by one additional condition at each end, since two are required. As illustrated in Fig. 2.28, in this instance the second condition would be zero bending moment.

Rotational Elastic Constraint

The rotational elastic constraint corresponds to a spring reaction moment that is equated to the bending moment. This is illustrated in Fig. 2.29. At the left end

$$EI\frac{\partial^2 v}{\partial x^2}(0, t) = k\frac{\partial v}{\partial x}(0, t) \qquad \rightarrow \qquad EIX''(0) = kX'(0). \tag{2.224}$$

At the right end

$$EI\frac{\partial^2 v}{\partial x^2}(\ell, t) = -k\frac{\partial v}{\partial x}(\ell, t) \qquad \rightarrow \qquad EIX''(\ell) = -kX'(\ell). \tag{2.225}$$

As in the previous case the signs differ and one more condition is required at each end.

Translational Inertia Constraint

The translational inertia constraint corresponds to the inertial reaction force associated with the translational acceleration of a rigid body or particle of mass m_c attached to the

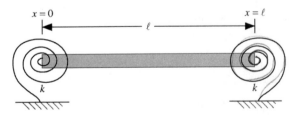

Figure 2.29   Schematic of rotational spring end conditions.

Figure 2.30   Schematic of translational mass end conditions.

end of the beam, as illustrated in Fig. 2.30. This force is equated to the shear. At the left end

$$EI\frac{\partial^3 v}{\partial x^3}(0, t) = -m_c\frac{\partial^2 v}{\partial t^2}(0, t). \tag{2.226}$$

From the previously given general solution

$$\frac{\partial^2 v}{\partial t^2} = X(x)\ddot{Y}(t) = -\omega^2 X(x)Y(t) = -\omega^2 v = -a^4\alpha^4 v, \tag{2.227}$$

and so the boundary condition can now be written as

$$EI\frac{\partial^3 v}{\partial x^3}(0, t) = m_c a^4\alpha^4 v(0, t) \qquad \rightarrow \qquad EIX'''(0) = m_c a^4\alpha^4 X(0). \tag{2.228}$$

At the right end the sign is changed as the result of the shear convention to yield

$$EIX'''(\ell) = -m_c a^4\alpha^4 X(\ell). \tag{2.229}$$

The above conditions must be augmented by one additional condition at each end, since two are required. As illustrated in Fig. 2.30, in this instance the second condition would be zero bending moment.

Rotational Inertia Constraint

The rotational inertia constraint corresponds to the inertial reaction moment associated with the rotational acceleration of a rigid body attached to the end of the beam, as shown in Fig. 2.31. This moment is equated to the bending moment. At the left end

$$EI\frac{\partial^2 v}{\partial x^2}(0, t) = I_c\frac{\partial^3 v}{\partial x\partial^2 t}(0, t) = -I_c a^4\alpha^4\frac{\partial v}{\partial x}(0, t), \tag{2.230}$$

and so the boundary condition can now be written as

$$EIX''(0) = -I_c a^4\alpha^4 X'(0). \tag{2.231}$$

At the right end the sign is changed as the result of the bending moment sign convention to yield

$$EIX''(\ell) = I_c a^4\alpha^4 X'(\ell). \tag{2.232}$$

We note here that when the attached mass is idealized as a particle, then $I_c = 0$; and the moment boundary condition reduces to be the same as indicated above for the translational mass, that is, the bending moment is zero.

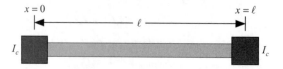

Figure 2.31   Schematic of rotational inertia end conditions.

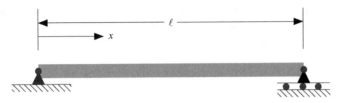

Figure 2.32   Schematic of pinned–pinned beam.

2.3.4   *Example Solutions for Mode Shapes and Frequencies*

In this section we consider several examples of the calculation of natural frequencies and mode shapes of vibrating beams in bending. One of the simplest cases is the pinned–pinned case, with which we begin. It is one of the few cases for beams in bending for which a numerical solution of the characteristic equation is *not* required. Next we treat the important clamped–free case, followed by the case of a hinged–free beam with a rotational restraint about the hinge. Finally, we consider the free–free case, illustrating the concept of the rigid-body mode.

Example 8: Solution for Pinned–Pinned Beam

Consider the pinned–pinned beam as shown in Fig. 2.32. The horizontal rollers at the right end indicate that the resultant axial force in the beam is zero. The boundary conditions reduce to conditions on X given by

$$X(0) = X''(0) = X(\ell) = X''(\ell) = 0. \tag{2.233}$$

Substituting the first two of the boundary conditions into the general solution as found in Eqs. (2.220), one finds that

$$\begin{aligned} D_2 + D_4 &= 0, \\ \alpha^2(-D_2 + D_4) &= 0. \end{aligned} \tag{2.234}$$

The constant α cannot be zero, because the form of the solution would change to a cubic polynomial, and the boundary conditions for this case do not yield a nontrivial solution of that form. Therefore, $D_2 = D_4 = 0$, and the solution for X becomes

$$X(x) = D_1 \sin(\alpha x) + D_3 \sinh(\alpha x). \tag{2.235}$$

Using the last two boundary conditions, one obtains a set of homogeneous algebraic equations in D_1 and D_3:

$$\begin{bmatrix} \sin(\alpha\ell) & \sinh(\alpha\ell) \\ -\sin(\alpha\ell) & \sinh(\alpha\ell) \end{bmatrix} \begin{Bmatrix} D_1 \\ D_3 \end{Bmatrix} = \begin{Bmatrix} 0 \\ 0 \end{Bmatrix}. \tag{2.236}$$

A nontrivial solution can only exist if the determinant of the coefficients is equal to zero, and so

$$2 \sin(\alpha\ell) \sinh(\alpha\ell) = 0. \tag{2.237}$$

Since $\alpha \neq 0$, we know that the only way this characteristic equation can be satisfied is for

$$\sin(\alpha\ell) = 0, \tag{2.238}$$

which has a denumerably infinite set of roots given by

$$\alpha_i = \frac{i\pi}{\ell} \qquad (i = 1, 2, \ldots). \tag{2.239}$$

Although this is the same set of eigenvalues that we found for the string problem, the relationship to the frequency is quite different. The frequencies are

$$\omega_i^2 = \alpha_i^4 a^4 = \frac{EI\alpha_i^4}{m},$$
(2.240)

so that

$$\omega_i = \alpha_i^2 \sqrt{\frac{EI}{m}} = \left(\frac{i\pi}{\ell}\right)^2 \sqrt{\frac{EI}{m}} = (i\pi)^2 \sqrt{\frac{EI}{m\ell^4}}.$$
(2.241)

As observed in the cases of the string and beam torsion, there is associated with the ith natural frequency a unique deformation shape called the mode shape (or eigenfunction). These mode shapes can be obtained from the spatially dependent portion of the solution by evaluating the function, $X_i(x)$, for any known value of α_i. To find X_i one substitutes any root back into either of Eqs. (2.236), recognizing that the constants D_1 and D_3 should now be written as D_{1_i} and D_{3_i}. Using the first of these equations along with the knowledge that $\sinh(\alpha_i \ell) \neq 0$, one finds that $D_{3_i} = 0$ and

$$X_i = D_{1_i} \sin\left(\frac{i\pi x}{\ell}\right) \qquad (i = 1, 2, \ldots),$$
(2.242)

where D_{1_i} can be anything. For example, choosing $D_{1_i} = 1$, one finds the mode shape to be

$$\phi_i = \sin\left(\frac{i\pi x}{\ell}\right) \qquad (i = 1, 2, \ldots),$$
(2.243)

which is the same mode shape as obtained earlier for the vibrating string.

Example 9: Solution for Clamped–Free Beam

Consider the clamped–free beam as shown in Fig. 2.33, the boundary conditions of which reduce to conditions on X given by

$$X(0) = X'(0) = X''(\ell) = X'''(\ell) = 0.$$
(2.244)

As with the previous example, one can show that this problem exhibits no nontrivial solution for the case of $\alpha = 0$. Thus, we use the form of the general solution in Eqs. (2.221) for which $\alpha \neq 0$. Along with the first two boundary conditions, this yields

$$X(0) = 0 \rightarrow E_3 = 0,$$
$$X'(0) = 0 \rightarrow E_1 = 0.$$
(2.245)

The remaining boundary conditions yield two homogeneous, algebraic equations, which may be reduced to the form

$$\begin{bmatrix} \sinh(\alpha\ell) + \sin(\alpha\ell) & \cosh(\alpha\ell) + \cos(\alpha\ell) \\ \cosh(\alpha\ell) + \cos(\alpha\ell) & \sinh(\alpha\ell) - \sin(\alpha\ell) \end{bmatrix} \begin{Bmatrix} E_2 \\ E_4 \end{Bmatrix} = \begin{Bmatrix} 0 \\ 0 \end{Bmatrix}.$$
(2.246)

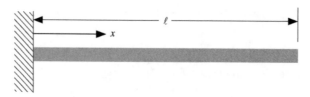

Figure 2.33 Schematic of clamped–free beam.

Table 2.1. *Values of* $\alpha_i \ell$, $(2i - 1)\pi/2$, *and* β_i *for* $i = 1, \ldots, 5$ *for the Clamped–Free Beam*

i	$\alpha_i \ell$	$(2i - 1)\pi/2$	β_i
1	1.87510	1.57080	0.734096
2	4.69409	4.71239	1.01847
3	7.85476	7.85398	0.999224
4	10.9955	10.9956	1.00003
5	14.1372	14.1372	0.999999

It can be verified by applying Cramer's method for their solution that a nontrivial solution only exists if the determinant of the coefficients is equal to zero. This is typical of all nontrivial solutions to homogeneous, linear, algebraic equations, and here yields

$$\sinh^2(\alpha\ell) - \sin^2(\alpha\ell) - [\cosh(\alpha\ell) + \cos(\alpha\ell)]^2 = 0, \tag{2.247}$$

or, noting the identities

$$\sin^2(\alpha\ell) + \cos^2(\alpha\ell) = 1,$$

$$\cosh^2(\alpha\ell) - \sinh^2(\alpha\ell) = 1 \tag{2.248}$$

one obtains the characteristic equation as simply

$$\cos(\alpha\ell)\cosh(\alpha\ell) + 1 = 0. \tag{2.249}$$

One cannot extract a closed-form exact solution for this transcendental equation. However, numerical solutions are easily obtained. Most numerical solution procedures require initial estimates of the solution in order for the procedure to converge. Since $\cosh(\alpha\ell)$ becomes large as its argument becomes large, one can argue that at least the larger roots will be close to those of $\cos(\alpha\ell) = 0$, or $\alpha_i\ell = (2i - 1)\pi/2$. Indeed, the use of these values as initial estimates yields a set of numerical values that approach the initial estimates ever more closely as i becomes large. The values of $\alpha_i\ell$ (dimensionless quantities) are listed in Table 2.1. To six places, all values of $\alpha_i\ell$ for $i \geq 5$ are equal to $(2i - 1)\pi/2$. The corresponding natural frequencies are given by

$$\omega_i = \alpha_i^2\sqrt{\frac{EI}{m}} = (\alpha_i\ell)^2\sqrt{\frac{EI}{m\ell^4}}. \tag{2.250}$$

To obtain the mode shapes, one substitutes the values in Table 2.1 back into either of Eqs. (2.246). The resulting equation for the ith mode has one arbitrary constant remaining (either E_{2i} or E_{4i} can be kept), which can be set equal to any number desired so as to normalize the resulting mode shape ϕ_i in some convenient way. For example, normalizing the solution by, $-E_{4i}$, which is equivalent to setting $E_{4i} = -1$, one can show that

$$\phi_i = \cosh(\alpha_i x) - \cos(\alpha_i x) - \beta_i[\sinh(\alpha_i x) - \sin(\alpha_i x)], \tag{2.251}$$

where

$$\beta_i = -\frac{E_{2i}}{E_{4i}} = \frac{\cosh(\alpha_i\ell) + \cos(\alpha_i\ell)}{\sinh(\alpha_i\ell) + \sin(\alpha_i\ell)} \tag{2.252}$$

and

$$\int_0^\ell \phi_i^2 dx = \ell,$$

$$\phi_i(\ell) = 2(-1)^{i+1}.$$

(2.253)

The values of β_i are also tabulated in Table 2.1. The first three mode shapes are depicted in Fig. 2.34. Note that the higher the mode number is, the more crossings of the zero displacement line occur.

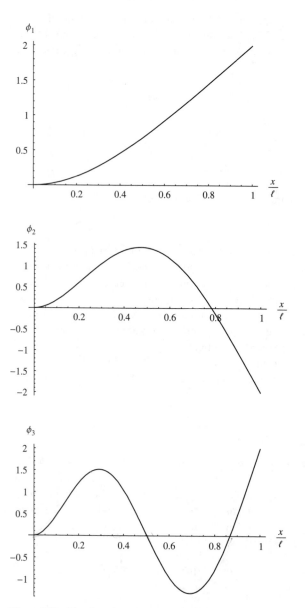

Figure 2.34   First three free-vibration mode shapes of a clamped–free beam in bending.

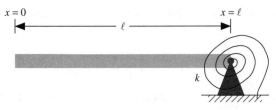

Figure 2.35   Schematic of spring-restrained, hinged-free beam.

Example 10: Solution for Spring-Restrained, Hinged–Free Beam

This sample problem for which modes of vibration will be determined is for a uniform beam that is hinged at the right-hand end and restrained there by a rotational spring with elastic constant $k = \kappa E I / \ell$. The left-hand end is free, as illustrated in Fig. 2.35. The boundary conditions for this case require that

$$X''(0) = 0,$$
$$X'''(0) = 0,$$
$$X(\ell) = 0,$$
$$E I X''(\ell) = -k X'(\ell) \text{ or } \ell X''(\ell) = -\kappa X'(\ell).$$

(2.254)

The spatially dependent portion of the general solution for beam bending dynamics will be used in the form of Eq. (2.221). The two conditions of zero bending moment and shear at $x = 0$ require that

$$X''(0) = 0 \rightarrow E_4 = 0,$$
$$X'''(0) = 0 \rightarrow E_2 = 0.$$

(2.255)

The third boundary condition, that of zero displacement at $x = \ell$, can now be indicated by

$$X(\ell) = E_1 \left[\sin(\alpha\ell) + \sinh(\alpha\ell)\right] + E_3 \left[\cos(\alpha\ell) + \cosh(\alpha\ell)\right] = 0.$$ (2.256)

The fourth boundary condition, a rotational elastic constraint at $x = \ell$, can be written as

$$\ell^2 X''(\ell) + \kappa \ell X'(\ell) = 0,$$

(2.257)

so that

$$(\alpha\ell)^2 \left\{ E_1 \left[-\sin(\alpha\ell) + \sinh(\alpha\ell)\right] + E_3 \left[-\cos(\alpha\ell) + \cosh(\alpha\ell)\right] \right\}$$
$$+ \kappa \alpha \ell \left\{ E_1 \left[\cos(\alpha\ell) + \cosh(\alpha\ell)\right] + E_3 \left[-\sin(\alpha\ell) + \sinh(\alpha\ell)\right] \right\} = 0.$$

(2.258)

This relation can be rearranged as

$$E_1 \left\{ \cos(\alpha\ell) + \cosh(\alpha\ell) + \frac{\alpha\ell}{\kappa} \left[-\sin(\alpha\ell) + \sinh(\alpha\ell)\right] \right\}$$
$$+ E_3 \left\{ -\sin(\alpha\ell) + \sinh(\alpha\ell) + \frac{\alpha\ell}{\kappa} \left[-\cos(\alpha\ell) + \cosh(\alpha\ell)\right] \right\} = 0.$$

(2.259)

The simultaneous solution of Eqs. (2.256) and (2.259) for nonzero values of E_1 and E_3 requires that the determinant of the 2×2 array formed by their coefficients must be zero. Setting the determinant formed from Eqs. (2.256) and (2.259) above to zero, one finds

$$[\sin(\alpha\ell) + \sinh(\alpha\ell)] \left\{ \sin(\alpha\ell) - \sinh(\alpha\ell) + \frac{\alpha\ell}{\kappa} [\cos(\alpha\ell) - \cosh(\alpha\ell)] \right\}$$

$$+ [\cos(\alpha\ell) + \cosh(\alpha\ell)] \left\{ \cos(\alpha\ell) + \cosh(\alpha\ell) + \frac{\alpha\ell}{\kappa} [-\sin(\alpha\ell) + \sinh(\alpha\ell)] \right\} = 0.$$

$$(2.260)$$

After executing the indicated multiplications and applying the identities of Eqs. (2.248), the relation becomes

$$\left(\frac{\alpha\ell}{\kappa} \right) [\sin(\alpha\ell)\cosh(\alpha\ell) - \cos(\alpha\ell)\sinh(\alpha\ell)] = 1 + \cos(\alpha\ell)\cosh(\alpha\ell). \quad (2.261)$$

This is the characteristic equation that can be solved for a denumerably infinite set of $\alpha\ell$s (the eigenvalues) for any specified value of κ. For specified finite and nonzero values of κ, a solution set can be obtained by numerical iteration on the unknown parameter $\alpha\ell$. The eigenvalues denoted by $\alpha_i \ell$ (for $i = 1, 2, \ldots$) can then be identified by satisfaction of the characteristic equation. In the limit of an infinite value of κ, one finds eigenvalues in agreement with the clamped–free case, as expected. In the limit of zero κ, a rigid-body mode can be shown to exist (see the next example for a problem that has rigid-body modes). For specified values of m, EI, ℓ, and the stiffness parameter κ, the eigenvalues can be used to determine the natural frequencies as

$$\omega_i = \alpha_i^2 \sqrt{\frac{EI}{m}} = (\alpha_i \ell)^2 \sqrt{\frac{EI}{m\ell^4}} \qquad (i = 1, 2, \ldots), \qquad (2.262)$$

and the ith mode shape can be defined as

$$\phi_i(x) = \frac{X_i(x)}{E_{1i}}$$

$$= \sin(\alpha_i x) + \sinh(\alpha_i x) + \beta_i [\cos(\alpha_i x) + \cosh(\alpha_i x)]. \qquad (2.263)$$

The modal parameter $\beta_i = E_{3i}/E_{1i}$ can be obtained from the zero-displacement boundary condition at $x = \ell$, Eq. (2.256). When evaluated for the ith mode, β_i becomes

$$\beta_i = \frac{E_{3i}}{E_{1i}} = -\frac{\sin(\alpha_i \ell) + \sinh(\alpha_i \ell)}{\cos(\alpha_i \ell) + \cosh(\alpha_i \ell)}, \qquad (2.264)$$

numerical values of which can be found once $\alpha_i \ell$ is known for specific values of κ.

A sample set of numerical results is shown in Fig. 2.36. In the first three figures, the first three mode shapes are shown for the case in which $\kappa = 1$. Figure 2.37 shows the variation of $\alpha_1 \ell$ versus κ, illustrating the fact that the frequencies of the higher modes are far less sensitive to the spring constant than that of the first mode. Indeed, the first mode frequency (proportional to the square of the smallest plotted quantity in Fig. 2.37) goes to zero as κ tends toward zero in the limit. This can be interpreted as the lowest-frequency mode transitioning to a rigid-body mode, which only exists when the spring constant is identically zero. In the limit as κ becomes infinite, in contrast, the eigenvalues tend toward those of the clamped–free beam, as expected. Indeed, as Fig. 2.38 shows when $\kappa = 50$, the mode shape starts to look more like that of a clamped–free beam (with the fixity being on the right end in this example).

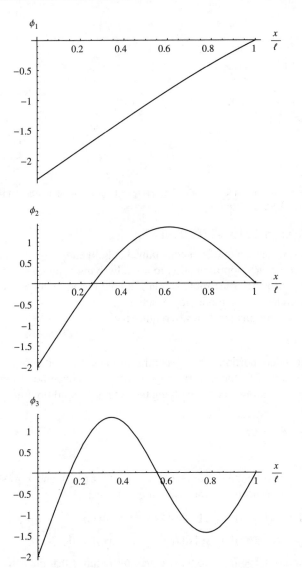

Figure 2.36 Mode shapes for first three modes of a spring-restrained, hinged–free beam in bending, $\kappa = 1$; $\omega_1 = (1.24792)^2 \sqrt{EI/(m\ell^4)}$, $\omega_2 = (4.03114)^2 \sqrt{EI/(m\ell^4)}$, and $\omega_3 = (7.13413)^2 \sqrt{EI/(m\ell^4)}$.

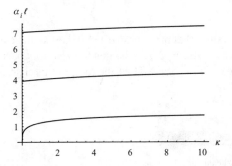

Figure 2.37 Variation of lowest eigenvalues $\alpha_i \ell$ versus dimensionless spring constant κ.

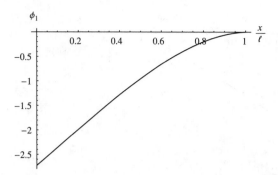

Figure 2.38 Mode shape for fundamental mode of the spring-restrained, hinged–free beam in bending; $\kappa = 50$ and $\omega_1 = (1.83929)^2 \sqrt{EI/(m\ell^4)}$.

Example 11: Solution for Free–Free Beam

The case of a uniform beam that is unconstrained at both ends, Fig. 2.39, may be considered as a crude first approximation to a freely flying vehicle. Their elastic and rigid dynamic properties are quite similar. In both instances these properties can be described in terms of a modal representation.

The boundary conditions for this case require that

$$X''(0) = X'''(0) = X''(\ell) = X'''(\ell) = 0. \tag{2.265}$$

The spatially dependent portion of the general solution to be used here will again involve the sums and differences of the trigonometric and hyperbolic functions. Two of the E_is can be eliminated by applying the boundary conditions at $x = 0$ so that

$$X''(0) = 0 \rightarrow E_4 = 0,$$
$$X'''(0) = 0 \rightarrow E_2 = 0. \tag{2.266}$$

The conditions of zero bending moment at $x = \ell$, $X''(\ell) = 0$, and zero shear at $x = \ell$, $X'''(\ell)$, yield, respectively, the following relations:

$$E_1 \left[-\sin(\alpha\ell) + \sinh(\alpha\ell)\right] + E_3 \left[-\cos(\alpha\ell) + \cosh(\alpha\ell)\right] = 0,$$
$$E_1 \left[-\cos(\alpha\ell) + \cosh(\alpha\ell)\right] + E_3 \left[\sin(\alpha\ell) + \sinh(\alpha\ell)\right] = 0. \tag{2.267}$$

Here again the nontrivial solution to these equations requires that the determinant of the E_1 and E_3 coefficients be zero. This relation becomes

$$\sinh^2(\alpha\ell) - \sin^2(\alpha\ell) - [\cosh(\alpha\ell) - \cos(\alpha\ell)]^2 = 0, \tag{2.268}$$

which simplifies to

$$\cos(\alpha\ell)\cosh(\alpha\ell) = 1. \tag{2.269}$$

For large $\alpha\ell$, the roots tend to values that make $\cos(\alpha\ell) = 0$. Unlike the clamped–free case, however, there is no root near $\pi/2$, and the first nonzero root occurs near $3\pi/2$. Indeed, the ith root is near $(2i + 1)\pi/2$. Thus, the roots of this characteristic

Figure 2.39 Schematic of free–free beam.

Table 2.2. *Values of $\alpha_i \ell$, $(2i + 1)\pi/2$, and β_i for*
$i = 1, \ldots, 5$ for the Free–Free Beam

i	$\alpha_i \ell$	$(2i + 1)\pi/2$	β_i
1	4.73004	4.71239	0.982502
2	7.85320	7.85398	1.00078
3	10.9956	10.9956	0.999966
4	14.1372	14.1372	1.00000
5	17.2788	17.2788	1.00000

equation can readily be computed numerically to yield the eigenvalues $\alpha_i \ell$ in Table 2.2. From these numerical values the natural frequencies can be found as

$$\omega_i = \alpha_i^2 \sqrt{\frac{EI}{m}}. \tag{2.270}$$

The mode shape associated with each eigenvalue can be defined as

$$\phi_i(x) = \frac{X_i(x)}{E_{3i}}$$
$$= \cos(\alpha_i x) + \cosh(\alpha_i x) - \beta_i[\sin(\alpha_i x) + \sinh(\alpha_i x)]. \tag{2.271}$$

The numerical value of the modal parameter $\beta_i = -E_{1i}/E_{3i}$, also tabulated in Table 2.2, can be obtained from either of the boundary conditions given above as Eqs. (2.267). Using the first of those equations as an example, one obtains

$$\beta_i = -\frac{E_{1i}}{E_{3i}} = \frac{\cosh(\alpha_i \ell) - \cos(\alpha_i \ell)}{\sinh(\alpha_i \ell) - \sin(\alpha_i \ell)}. \tag{2.272}$$

It can be shown that the second of Eqs. (2.267) would yield the same result by using the characteristic equation as an identity. The first three of these mode shapes are shown in Fig. 2.40.

In addition to these modal properties that can be used to describe the elastic behavior of the beam, there are also modal properties that describe the rigid behavior of the beam. These modes are associated with zero values of the separation constant α. It should be recalled that a similar result was obtained for torsional deflections of a free–free beam. When α is zero, the governing ordinary differential equations for beam bending, Eqs. (2.213), become

$$X'''' = 0, \qquad \ddot{Y} = 0. \tag{2.273}$$

The general solutions to these equations can be written as

$$X = \frac{b x^3}{6} + \frac{c x^2}{2} + d x + e,$$
$$Y = ft + g, \tag{2.274}$$

where the arbitrary constants, b through e, in the spatially dependent portion of the solution can be established from the boundary conditions. These conditions of zero bending moment and shear at the ends of the beam yield the following:

$$X''(0) = 0 \rightarrow c = 0,$$
$$X'''(0) = 0 \rightarrow b = 0,$$
$$X''(\ell) = 0 \rightarrow b\ell + c = 0, \tag{2.275}$$
$$X'''(\ell) = 0 \rightarrow b = 0.$$

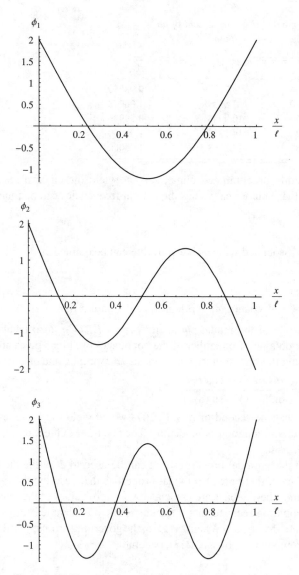

Figure 2.40   First three free-vibration elastic mode shapes of a free–free beam in bending.

It is apparent that all four boundary conditions can be satisfied with $b = c = 0$. Since no restrictions are placed on the constants d and e, they can be arbitrary. Thus, a general description of the solution in this case is

$$X = dx + e. \tag{2.276}$$

A very important characteristic of this solution is that no relationship has been established between d and e. Therefore, they can be presumed to represent two independent motions of the beam. As written above, e represents a rigid vertical translation of the beam since it is independent of x. The dx term, being linear in x, represents a rigid rotation of the beam about the left end. It can be shown that when the translation is of the mass centroid and the rotational motion is about the mass centroid the two motions are orthogonal with respect to each other and with respect to the elastic modes. This suggests that the modal representation for these rigid-body degrees of freedom can be described by

$$v_{\text{rigid}} = \sum_{i=-1}^{0} \phi_i(x)\xi_i(t), \tag{2.277}$$

where

$$\begin{aligned} \phi_{-1} &= 1 &&\text{and} &&\xi_{-1}(t) = \text{c.g. translation,} \\ \phi_0 &= x - \frac{\ell}{2} &&\text{and} &&\xi_0(t) = \text{rotation angle.} \end{aligned} \tag{2.278}$$

The time-dependent portion of the solution for these rigid-body type motions is seen to be aperiodic. This means that natural frequencies for both rigid-body modes are zero. The two arbitrary constants contained in $Y(t)$ can be evaluated from the initial rigid-body displacement and velocity associated with the translation and rotation. Thus, the complete solution for the free–free beam bending problem can now be written in terms of all of its modes as

$$v = \sum_{i=-1}^{\infty} \phi_i(x)\xi_i(t). \tag{2.279}$$

This example provides a convenient vehicle for further discussion of symmetry. It has already been noted in the case of a vibrating string that systems exhibiting geometric symmetry have two distinct types of mode shapes, namely, those that are symmetric about the midpoint and those that are antisymmetric about the midpoint. As can be seen in the results, this is indeed true for the modes of the free–free beam. In particular, the rigid-body translation mode and the first and third elastic modes are clearly symmetric about the mid-point of the beam, whereas the rigid-body rotation mode and the second elastic mode are antisymmetric about the midpoint (see Fig. 2.40).

This observation suggests that the symmetric mode shapes could be obtained by calculating the mode shapes of a beam that is half the length of the original beam and that has the sliding condition at one end and is free at the other. Similarly, the antisymmetric modes could be obtained by calculating the mode shapes of a beam with half the length of the original beam and that has one end pinned and the other free. It should also be evident that a symmetric aircraft with high-aspect-ratio wings, modeled as beams and attached to a rigid-body fuselage, could be represented in terms of the symmetric and antisymmetric modes of the combined body and wing system in a similar way. That is, one may model the whole system by only considering one wing attached to a rigid body with half the mass and half the rotational inertia with appropriate boundary conditions.

2.4   Approximate Solution Techniques

There are several popular methods that make use of a set of modes or other functions to approximate the dynamic behavior of systems. In this section, without going into details on the theories associated with this subject, we will illustrate within the framework already established how one can use a truncated set of modes or other set of functions to obtain an approximate solution. Details of the theories behind modal approximation methods may be found in texts that treat structural dynamics at the graduate level. The two main approaches are Galerkin's method, applied to ordinary or partial differential equations, and the Ritz method, applied to the principle of virtual work. These two methods yield identical results in certain situations. Thus, if time is limited it would only be necessary to discuss one of the two methods to give the student an introduction to the method and an appreciation of results that can be obtained this way. The Ritz method is to be preferred in the present context because of the ease with which it can be presented within the framework of Lagrange's equations. Nevertheless, both of these methods will be presented at a level suitable for undergraduate students.

2.4.1 *The Ritz Method*

Building on the earlier treatment, we start with Lagrange's equations, given by

$$\frac{d}{dt}\left(\frac{\partial L}{\partial \dot{\xi}_i}\right) - \frac{\partial L}{\partial \xi_i} = \Xi_i \qquad (i = 1, 2, \ldots, n), \tag{2.280}$$

where the Lagrangean, $L = K - P$, the total kinetic energy is K, the total potential energy is P, n is the number of generalized coordinates retained, the generalized coordinates are ξ_i, and Ξ_i is the generalized force. Although it can be helpful, as we shall see below, it is not necessary to make use of potential energy, which can only account for conservative forces. The generalized force, however, can be used to include the effects of *any* loads. So as not to count the same physical effects more than once, the generalized force should only include those forces that are not accounted for in the potential energy. Generalized forces stem from virtual work, which can be written as

$$\overline{\delta W} = \sum_{i=1}^{n} \Xi_i \delta \xi_i, \tag{2.281}$$

where $\delta \xi_i$ is an arbitrary increment in the ith generalized coordinate.

Let us consider a beam in bending as an example. The total kinetic energy must include that of the beam as well as of any attached particles or rigid bodies. The contribution of the beam is

$$K_{\text{beam}} = \frac{1}{2} \int_0^\ell m \left(\frac{\partial v}{\partial t}\right)^2 dx, \tag{2.282}$$

where m is the mass per unit length of the beam. The total potential energy $P = U + V$ comprises the internal strain energy of the beam, denoted by U, plus any additional potential energy, V, attributed to gravity, springs attached to the beam, or applied static loads. All other loads, such as aerodynamic loads, damping, follower forces, etc., must be accounted for in Ξ_i.

The strain energy for a beam in bending is given by

$$U = \frac{1}{2} \int_0^\ell EI \left(\frac{\partial^2 v}{\partial x^2}\right)^2 dx. \tag{2.283}$$

The expression for V varies depending on the problem being addressed, as does the virtual work of all forces other than those accounted for in V. The virtual work can be written as

$$\overline{\delta W} = \int_0^\ell f(x, t) \delta v(x, t) \, dx, \tag{2.284}$$

where δv is an increment of v in which time is held fixed and $f(x, t)$ is the distributed force per unit length and is positive in the direction of positive v.

To apply the Ritz method, one needs to express P, K, and $\overline{\delta W}$ in terms of a series of functions with one or more terms. For a beam in bending, this means that

$$v(x, t) = \sum_{i=1}^{n} \xi_i(t) \phi_i(x). \tag{2.285}$$

There are several characteristics that these "basis functions" ϕ_i must possess:

1. Each function at least must satisfy all boundary conditions on displacement and rotation (often called the "geometric" boundary conditions). It is not necessary that they satisfy the force and moment boundary conditions, but satisfaction of them may improve accuracy. However, it is not easy in general to find functions that satisfy all boundary conditions.
2. Each function must be continuous and p times differentiable, where p is the order of the highest spatial derivative in the Lagrangean. The pth derivative of at least one function must be nonzero. Here, from Eq. (2.283), $p = 2$.
3. If more than one function is used, they must be chosen from a set of functions that is complete. This means that any function on the interval $0 \leq x \leq \ell$ and having the same boundary conditions as the problem under consideration can be expressed to any degree of accuracy whatsoever as a linear combination of the functions in the set. Examples of complete sets of functions on the interval $0 \leq x \leq \ell$ include

$$1, x, x^2, \ldots,$$

$$\sin\left(\frac{\pi x}{\ell}\right), \sin\left(\frac{2\pi x}{\ell}\right), \sin\left(\frac{3\pi x}{\ell}\right), \ldots, \text{or}$$

a set of mode shapes for *any* problem.

Completeness also implies that there can be no missing terms between the lowest and highest ones used in any series.

4. The set of functions must be linearly independent. This means that

$$\sum_{i=0}^{n} a_i \phi_i(x) = 0 \quad \Rightarrow \quad a_i = 0 \text{ for all } i. \tag{2.286}$$

A set of functions that satisfies all these criteria is said to be "admissible."

By use of the series approximation, we have reduced a problem with an infinite number of degrees of freedom to one with n degrees of freedom. Instead of being governed by a partial differential equation, the behavior of this system is now defined by n second-order, ordinary differential equations in time. This reduction from a continuous system modeled by a partial differential equation with an infinite number of degrees of freedom to one described by a finite number of ordinary differential equations in time is sometimes called spatial discretization. The number n is usually increased until convergence is obtained. (It should be noted that if inertial forces are not considered, so that the kinetic energy is identically zero, then a system described by an ordinary differential equation in a single spatial variable is reduced by the Ritz method to one described by n algebraic equations.)

Now, let us illustrate how the approximating functions are actually used. Let ϕ_i, $i = 1$, $2, \ldots, \infty$, be a complete set of p-times differentiable, linearly independent functions that satisfy the displacement and rotation boundary conditions. Thus, U can be written as

$$U = \frac{1}{2} \sum_{i=1}^{n} \sum_{j=1}^{n} \xi_i \xi_j \int_0^{\ell} EI \phi_i'' \phi_j'' dx. \tag{2.287}$$

The contributions of any springs that restrain the structure, as well as conservative loads, must be added to obtain the full potential energy P.

The kinetic energy of the beam is

$$K_{\text{beam}} = \frac{1}{2} \sum_{i=1}^{n} \sum_{j=1}^{n} \dot{\xi}_i \dot{\xi}_j \int_0^{\ell} m \phi_i \phi_j dx. \tag{2.288}$$

Contributions of any additional particles and rigid bodies must be added to obtain the complete kinetic energy K.

The virtual work term must account for distributed and concentrated forces resulting from all other sources, such as damping, aerodynamics, etc. This can be written as

$$\overline{\delta W} = \sum_{i=1}^{n} \delta\xi_i \left[\int_0^\ell f(x, t)\phi_i dx + F_c(x_0, t)\phi_i(x_0) \right], \tag{2.289}$$

where x_0 is a value of x at which a concentrated force is located. Here the first term accounts for a distributed force $f(x, t)$ on the interior of the beam, and the second term accounts for a concentrated force on the interior (see Eq. 2.112). In aeroelasticity, the loads $f(x, t)$ and $F_c(x_0, t)$ may depend on the displacement in some complicated manner.

The integrands in the above quantities all involve the basis functions and their derivatives over the length of the beam. It should be noted that these integrals involve only known quantities and often can be evaluated analytically. Sometimes they are too complicated to undertake analytically, however, and they must be evaluated numerically. Numerical evaluation is often facilitated by nondimensionalization. Symbolic computation tools such as Mathematica and Maple may be quite helpful in both situations.

With all such things considered, the equations of motion can be written in a form that is quite common, that is,

$$[M]\{\ddot{\xi}\} + [C]\{\dot{\xi}\} + [K]\{\xi\} = \{F\}, \tag{2.290}$$

where $\{\xi\}$ is a column matrix of the generalized coordinates, $\{F\}$ is a column matrix of the generalized force terms that do not depend on ξ_i, (̇) is the time derivative of ( ), $[M]$ is the mass matrix, $[C]$ is the gyroscopic/damping matrix, and $[K]$ is the stiffness matrix. The most important contribution to $[M]$ is from the kinetic energy, and this contribution is symmetric. The most important contribution to $[K]$ is from the strain energy of the structure and potential energy of any springs that restrain the motion of the structure. There can be contributions to all terms in the equations of motion from kinetic energy and virtual work. For example, there are contributions from kinetic energy to $[C]$ and $[K]$ when there is a rotating coordinate system. Damping makes contributions to $[C]$ through the virtual work. Finally, because aerodynamic loads in general depend on the displacement and its time derivatives, aeroelastic analyses may contain terms in $[M]$, $[C]$, and $[K]$ that stem from aerodynamic loads.

An interesting special case of this method occurs when the system is conservatively loaded. The resulting method is usually referred to as the Rayleigh–Ritz method, and there are many theorems that can be proved about the convergence of approximations to the natural frequency. Indeed, one of the most powerful of such theorems states that the approximate natural frequencies are always upper bounds, and another states that adding more terms to a given series always lowers the approximate natural frequencies (i.e., making them closer to the exact values).

A further specialized case is the simplest approximation, in which only one term is used. Then, an approximate expression for the lowest natural frequency can be written as a ratio called the Rayleigh quotient. This simplest special case is of more than merely academic interest. It is not at all uncommon that a rough estimate of the lowest natural frequency is needed early in the design of flexible structures.

Example 12: The Ritz Method Using Clamped–Free Modes

In the first example we consider a uniform, clamped–free beam that we modify by adding a tip mass of mass $\mu m\ell$. The exact solution can be easily obtained for this modified problem using the methodology spelled out earlier. However, it is desired here to illustrate the Ritz method, and we already have calculated the modes for a clamped–free beam (i.e., without a tip mass) in Section 2.3.4. These mode shapes are solutions of an eigenvalue problem, and so, provided we do not skip any modes between the lowest and highest mode number that we use, this set is automatically complete. The set is also orthogonal and therefore linearly independent. Of course, these modes automatically satisfy the boundary conditions on displacement and rotation for our modified problem (since they are the same as for the clamped–free beam), and they are infinitely differentiable. Hence, they are admissible functions for the modified problem. Moreover, they satisfy the zero moment boundary condition at the free end, which is a boundary condition for our modified problem. However, because of the presence of the tip mass in the modified problem, the shear force, which the reader will recall is proportional to the third derivative of the displacement, does not vanish as it does for the clamped–free mode shapes.

The strain energy becomes

$$U = \frac{1}{2} \sum_{i=1}^{n} \sum_{j=1}^{n} \xi_i \xi_j \int_0^\ell EI \phi_i'' \phi_j'' dx. \tag{2.291}$$

Substituting the mode shapes of Eq. (2.251) into Eq. (2.291) and taking advantage of orthogonality, one can simplify it to

$$U = \frac{\ell EI}{2} \sum_{i=1}^{n} \xi_i^2 \alpha_i^4, \tag{2.292}$$

where α_i is the set of constants in Table 2.1. Similarly, accounting for the tip mass, the kinetic energy of which is

$$K_{\text{tip mass}} = \frac{1}{2} \mu m \ell \left[\frac{\partial v}{\partial t}(\ell, t) \right]^2$$
$$= \frac{1}{2} \mu m \ell \sum_{i=1}^{n} \sum_{j=1}^{n} \dot{\xi}_i \dot{\xi}_j \phi_i(\ell) \phi_j(\ell), \tag{2.293}$$

one obtains the total kinetic energy as

$$K = \frac{1}{2} \sum_{i=1}^{n} \sum_{j=1}^{n} \dot{\xi}_i \dot{\xi}_j \left[\int_0^\ell m \phi_i \phi_j dx + \mu m \ell \phi_i(\ell) \phi_j(\ell) \right]. \tag{2.294}$$

With the use of the mode shapes of Eq. (2.251), one finds that $\phi_i(\ell) = 2(-1)^{i+1}$ and so the kinetic energy simplifies to

$$K = \frac{m\ell}{2} \sum_{i=1}^{n} \sum_{j=1}^{n} \dot{\xi}_i \dot{\xi}_j [\delta_{ij} + 4\mu(-1)^{i+j}], \tag{2.295}$$

where the Kronecker symbol $\delta_{ij} = 1$ for $i = j$ and $\delta_{ij} = 0$ for $i \neq j$. For free vibration, there are no additional forces. Thus, Lagrange's equations can be now written in matrix form as

$$[M]\{\ddot{\xi}\} + [K]\{\xi\} = 0, \tag{2.296}$$

where $[K]$ is a diagonal matrix with the diagonal elements given by

$$K_{ii} = EI \ell \alpha_i^4 \qquad (i = 1, 2, \ldots, n) \tag{2.297}$$

Table 2.3. *Approximate Values of* $\omega_1\sqrt{m\ell^4/(EI)}$ *for
Clamped–Free Beam with Tip Mass of* $\mu m\ell$ *Using* n
Clamped–Free Modes of Section 2.3.4, Eq. (2.251)

n	μ		
	1	10	100
1	1.57241	0.549109	0.175581
2	1.55964	0.542566	0.173398
3	1.55803	0.541748	0.173126
4	1.55761	0.541536	0.173055
5	1.55746	0.541458	0.173029
Exact	1.55730	0.541375	0.173001

and $[M]$ is a symmetric matrix with elements given by

$$M_{ij} = m\ell[\delta_{ij} + 4\mu(-1)^{i+j}] \qquad (i, j = 1, 2, \ldots, n). \tag{2.298}$$

Assuming $\xi = \bar{\xi}\exp(i\omega t)$, one can write Eq. (2.296) as an eigenvalue problem
of the form

$$[[K] - \omega^2[M]]\{\bar{\xi}\} = 0. \tag{2.299}$$

Results for the first modal frequency are shown in Table 2.3 and compared therein
with the exact solution. As one can see, the approximate solution agrees with the
exact solution to within engineering accuracy with only two terms. By way of contrast, results for the second modal frequency are shown in Table 2.4. These results
are not nearly as accurate. Results for the higher modes, not shown, are less accurate
still. This is one of the problems with modal approximation methods; fortunately,
however, aeroelasticians and structural dynamicists are frequently interested only
in the lower-frequency modes. Note that the one-term approximation (i.e., the
Rayleigh quotient) is within 1.1% for all values of μ displayed.

Example 13: The Ritz Method Using a Simple Power Series

As an alternative to using the mode shapes of a closely related problem, let us
repeat the solution of the above using a simple power series to construct a series
of functions ϕ_i. Since the moment vanishes at the free end where $x = \ell$, one can
make the second derivative of all terms proportional to $\ell - x$. To get a complete
series, one can multiply this term by a complete power series $1, x, x^2$, etc. Thus,

Table 2.4. *Approximate Values of* $\omega_2\sqrt{m\ell^4/(EI)}$ *for
Clamped–Free Beam with Tip Mass of* $\mu m\ell$ *Using* n
Clamped–Free Modes of Section 2.3.4, Eq. (2.251)

n	μ		
	1	10	100
2	16.5580	15.8657	15.7867
3	16.3437	15.6191	15.5367
4	16.2902	15.5576	15.4744
5	16.2708	15.5353	15.4518
Exact	16.2501	15.5115	15.4277

Table 2.5. *Approximate Values of* $\omega_1 \sqrt{m\ell^4/(EI)}$ *for Clamped–Free Beam with Tip Mass of* $\mu m\ell$ *Using n Polynomial Functions*

n	μ		
	1	10	100
1	1.55812	0.541379	0.173001
2	1.55733	0.541375	0.173001
3	1.55730	0.541375	0.173001
4	1.55730	0.541375	0.173001
5	1.55730	0.541375	0.173001
Exact	1.55730	0.541375	0.173001

one may then write the second derivative of the ith function as

$$\phi_i'' = \frac{1}{\ell^2}\left(1 - \frac{x}{\ell}\right)\left(\frac{x}{\ell}\right)^{i-1}. \tag{2.300}$$

With the boundary conditions on displacement and rotation being $\phi_i(0) = \phi_i'(0) = 0$, one can then integrate to find an expression for the ith function as

$$\phi_i = \frac{\left(\frac{x}{\ell}\right)^{i+1}\left[2 + i - i\left(\frac{x}{\ell}\right)\right]}{i\,(1+i)\,(2+i)}. \tag{2.301}$$

Since the chosen admissible functions have non-zero third derivatives at the tip, they offer the possibility of satisfying the non-zero shear condition in combination with each other. Such admissible functions are sometimes called *quasi-comparison* functions.

In this case, the stiffness matrix becomes

$$K_{ij} = \frac{2EI}{\ell^3\,(i + j - 1)\,(i + j)\,(1 + i + j)} \qquad (i, j = 1, 2, \dots, n) \tag{2.302}$$

and the mass matrix is

$$M_{ij} = \frac{2m\ell\left[3(i^2 + j^2) + 7ij + 23(i + j) + 40\right]}{ij\,(i + 1)\,(i + 2)\,(j + 1)\,(j + 2)\,(i + j + 3)\,(i + j + 4)\,(i + j + 5)}$$
$$+ \frac{4\,\mu m\ell}{ij\,(i + 1)\,(i + 2)\,(j + 1)\,(j + 2)} \qquad (i, j = 1, 2, \dots, n). \tag{2.303}$$

Results from this calculation are given in Tables 2.5 and 2.6 for the first two modes. It is clear that these results are much better than those obtained with the

Table 2.6. *Approximate Values of* $\omega_2 \sqrt{m\ell^4/(EI)}$ *for Clamped–Free Beam with Tip Mass of* $\mu m\ell$ *Using n Polynomial Functions*

n	μ		
	1	10	100
2	16.2853	15.5443	15.4605
3	16.2841	15.5371	15.4524
4	16.2505	15.5119	15.4280
5	16.2501	15.5116	15.4277
Exact	16.2501	15.5115	15.4277

clamped–free beam modes. It is not unusual for polynomial functions to provide better results than those obtained with beam mode shapes. However, here it is worth noting that the beam mode shapes are at a disadvantage for this problem. Unlike the problem being solved (and the polynomials chosen), the beam mode shapes are constrained to have zero shear force at the free end and are thus not quasi-comparison functions for the problem with a tip mass. This one-term polynomial approximation (i.e., the Rayleigh quotient) is within 0.05%, which is exceptionally good given its simplicity.

It is sometimes suggested that the mode shapes of a closely related problem are, at least in some sense, superior to other approximate sets of functions. For example, in the first example we did see that the orthogonality of the modes used resulted in a diagonal stiffness matrix. That does provide a slight advantage in the ease of computing the eigenvalues. However, for the low-order problems of the sort we are discussing, that advantage is hardly noticeable. Indeed, symbolic computation tools such as Mathematica and Maple are capable of calculating the eigenvalues for problems of the size of this example in but a few seconds. Moreover, in some cases the simplicity of carrying out the integrals that result in approximate formulations is a more important factor in deciding what set of functions to use in carrying out the standard implementation of the Ritz method. Indeed, polynomial functions are generally much easier to deal with analytically than free-vibration modes such as those illustrated in Section 2.4.1, which frequently involve transcendental functions.

Alternatives to the standard Ritz method include the methods of Galerkin, finite elements, component mode synthesis, and influence coefficients. We shall touch on Galerkin's method in the next section. Detailed descriptions of the other approaches may be found in more advanced texts on structural dynamics and aeroelasticity. The finite element method is by far the most popular way of solving realistic structural dynamics and aeroelasticity problems in industry. The finite element method can be developed as a special case of the Ritz method, and it typically makes use of polynomial shape functions over each of the finite elements into which the original structure is broken. Finite element equations have the same structure as Eq. (2.290) but are typically of large order, with n being of the order of thousands to millions. What keeps the computational effort from being overly burdensome is that the matrices have a narrow-banded structure, which allows specialized software to be used in solving the equations of motion that takes advantage of this structure, reducing both memory and floating point operations and resulting in very significant computational advantages.

2.4.2   *Galerkin's Method*

Rather than making use of energy and Lagrange's equation as with the Ritz method, Galerkin's method starts with the partial differential equation of motion. Let us denote this equation by

$$\mathcal{L}[v(x, t)] = 0, \tag{2.304}$$

where $\mathcal{L}$ is an operator on the unknown function $v(x, t)$ with maximum spatial partial derivatives of the order q. For the structural dynamics problems we have addressed so far, the operator $\mathcal{L}$ is linear and $q = 2p$, where p is the maximum order of spatial partial derivative in the Lagrangean. It is important to note, however, that it is not true in general that $q = 2p$; indeed, one does not need to consider the Lagrangean at all with this method.

To apply Galerkin's method, one needs to express $v(x, t)$, and hence the operator $\mathcal{L}$, in terms of a series of functions with one or more terms. For a beam in bending, for example, this means as before that

$$v(x, t) = \sum_{j=1}^{n} \xi_j(t)\phi_j(x). \tag{2.305}$$

Relative to the basis functions used in the Ritz method, the characteristics that these functions ϕ_i must possess for use in Galerkin's method are more stringent:

1. Each function must satisfy all boundary conditions. Note that it is not easy in general to find functions that satisfy all boundary conditions.
2. Each function must be at least q times differentiable. The qth derivative of at least one function must be nonzero.
3. If more than one function is used, they must be chosen from a set of functions that is complete.
4. The set of functions must be linearly independent.

Functions that satisfy all these criteria are said to be "comparison functions." The original partial differential equation is then multiplied by ϕ_i and integrated over the domain of the independent variable (say, $0 \leq x \leq \ell$). Thus, a set of n ordinary differential equations is obtained from the original partial differential equation. (It should be noted that if the original equation is an ordinary differential equation in x, then Galerkin's method yields n algebraic equations.)

Let us consider a beam in bending as an example. The equation of motion can be written as in Eq. (2.208), with a slight change in notation, as

$$\frac{\partial^2}{\partial x^2}\left(EI\frac{\partial^2 v}{\partial x^2}\right) + m\frac{\partial^2 v}{\partial t^2} - f(x, t) = 0, \tag{2.306}$$

where EI is the flexural rigidity, m is the mass per unit length, and the boundary conditions and loading term $f(x, t)$ must reflect any attached particles or rigid bodies. In aeroelasticity, the loads $f(x, t)$ may depend on the displacement in some complicated manner.

With all the ingredients as described above considered, the discretized equations of motion can be written in the same form as with the Ritz method, that is,

$$[M]\{\ddot{\xi}\} + [C]\{\dot{\xi}\} + [K]\{\xi\} = \{F\}, \tag{2.307}$$

where $\{\xi\}$ is a column matrix of the generalized coordinates, $\{F\}$ is a column matrix of the generalized force terms that do not depend on ξ_i, ($\dot{\,}$) is the time derivative of ( ), $[M]$ is the mass matrix, $[C]$ is the gyroscopic/damping matrix, and $[K]$ is the stiffness matrix. As before, inertial forces contribute to $[M]$, and there are contributions from the inertial forces to $[C]$ and $[K]$ when there is a rotating coordinate system, and damping also contributes to $[C]$. Finally, because aeroelastic loads in general depend on the displacement and its time derivatives, aerodynamics can contribute terms to $[M]$, $[C]$, and $[K]$.

Example 14: Galerkin's Method for a Beam in Bending

Now, let us illustrate how the approximating functions are actually used. Let ϕ_i, $i = 1, 2, \ldots, \infty$, be a complete set of q-times differentiable, linearly independent functions that satisfy all the boundary conditions. Substituting Eq. (2.305) into Eq. (2.306), multiplying by $\phi_i(x)$, and integrating over x from 0 to ℓ, one obtains

$$\int_0^\ell \phi_i\left[\sum_{j=1}^n \xi_j(EI\phi_j'')'' + \sum_{j=1}^n \ddot{\xi}_j m\phi_j - f(x, t)\right] dx = 0 \qquad (i = 1, 2, \ldots, n).$$

$$\tag{2.308}$$

After reversing the order of integration and summation and integrating the first term by parts, taking into account the boundary conditions, this equation becomes

$$\sum_{j=1}^{n} \left(\xi_j \int_0^\ell EI\phi_i'' \phi_j'' dx + \ddot{\xi}_j \int_0^\ell m\phi_i \phi_j dx \right)$$

$$- \int_0^\ell f\phi_i dx = 0 \qquad (i = 1, 2, \ldots, n). \qquad (2.309)$$

When one compares the first two terms with the earlier derivation by the Ritz method, one sees the close relationship between these approaches. Indeed, if the starting partial differential equation is derivable from energy, which implies that $q = 2p$, and the same modal functions ϕ_i are used in both cases, the resulting discretized equations are the same.

Considering the clamped–free case, for example, one can develop a set of comparison functions by starting with

$$\phi_i'' = \frac{1}{\ell^2} \left(1 - \frac{x}{\ell} \right)^2 \left(\frac{x}{\ell} \right)^{i-1}. \qquad (2.310)$$

With the boundary conditions on displacement and rotation being $\phi_i(0) = \phi_i'(0) = 0$, one can then integrate to find an expression for the ith function as

$$\phi_i = \frac{\left(\frac{x}{\ell} \right)^{1+i} \left\{ 6 + i^2 \left(1 - \frac{x}{\ell} \right)^2 + i \left[5 - \frac{6x}{\ell} + \left(\frac{x}{\ell} \right)^2 \right] \right\}}{i(1+i)(2+i)(3+i)} \qquad (2.311)$$

Elements of the stiffness matrix can be found as

$$K_{ij} = \int_0^\ell EI\phi_i'' \phi_j'' dx$$

$$= \frac{24EI}{\ell^3 (i+j-1)(i+j)(1+i+j)(2+i+j)(3+i+j)}. \qquad (2.312)$$

Similarly, the elements of the mass matrix can be found as

$$M_{ij} = \int_0^\ell m\phi_i \phi_j dx$$

$$= \frac{m\ell p_1}{p_2}, \qquad (2.313)$$

where

$$p_1 = 30240 + 28512(i+j) + 9672(i^2+j^2) + 1392(i^3+j^3) + 72(i^4+j^4)$$
$$+ 20040ij + 4520(i^2j + ij^2) + 320(i^3j + ij^3) + 520i^2 j^2,$$

$$p_2 = i(1+i)(2+i)(3+i)j(1+j)(2+j)(3+j)(3+i+j)$$
$$\times (4+i+j)(5+i+j)(6+i+j)(7+i+j). \qquad (2.314)$$

The fact that the governing equation is derivable from energy is reflected in the symmetry of $[M]$ and $[K]$. Results for free vibration (i.e., with $f = 0$) are given in Table 2.7. As with the Ritz method, one sees monotonic convergence from above and accuracy comparable to that achieved via the Ritz method. However, unlike the Ritz method, one does not always obtain results for free-vibration problems that converge from above.

Table 2.7. *Approximate Values of* $\omega_i \sqrt{m\ell^4/(EI)}$ *for* $i = 1, 2,$
and 3, *for a Clamped–Free Beam Using n Polynomial Functions*

	Mode		
n	1	2	3
1	3.53009	—	—
2	3.51604	22.7125	—
3	3.51602	22.0354	66.2562
4	3.51602	22.0354	61.7675
5	3.51602	22.0345	61.7395
Exact	3.51602	22.0345	61.6972

Example 15: Galerkin's Method for a Beam in Bending Using an Alternative
Form of the Equation of Motion

Consider again a clamped–free beam. To obtain an alternative equation of motion
we integrate the equation of motion twice and use the zero shear and bending
moment boundary conditions to obtain an integro-partial differential equation

$$EI\frac{\partial^2 v}{\partial x^2} + \int_x^{\ell} (x - \zeta)\left[f(\zeta, t) - m\frac{\partial^2 v(\zeta, t)}{\partial t^2}\right]d\zeta = 0, \tag{2.315}$$

where ζ is a dummy variable. Although this equation of motion is somewhat more
complicated, it is only a second-order equation. Thus, it has only two boundary
conditions, which are zero displacement and slope at $x = 0$. Thus, a much simpler
set of comparison functions can be used, such as a simple power series $x^2, x^3, \ldots,$
that is,

$$\phi_i = x^{i+1} \qquad (i = 1, 2, \ldots, n). \tag{2.316}$$

We should not expect greater accuracy from this simple set of functions, but the
analytical effort is considerably less. Indeed, the elements of the stiffness matrix are

$$\begin{aligned} K_{ij} &= \int_0^{\ell} EI\phi_i\phi_j''dx \\ &= \frac{EIj(j+1)}{\ell(i+j+1)}, \end{aligned} \tag{2.317}$$

and the elements of the mass matrix are

$$\begin{aligned} M_{ij} &= \int_0^{\ell} \phi_i \int_x^{\ell} (\zeta - x)m\phi_j(\zeta)d\zeta \, dx \\ &= \frac{m\ell^3}{(2+i)(3+i)(5+i+j)}. \end{aligned} \tag{2.318}$$

Note that these matrices are not symmetric. Moreover, the results that are presented in Table 2.8 are not as accurate as those obtained before in Table 2.7, and
the convergence is not monotonic from above.

The partial differential equations derived earlier in this chapter for free vibration
of strings, beams in torsion, and beams in bending can be derived from energy-
based approaches, such as Hamilton's principle. (The use of Hamilton's principle is
beyond the scope of this text, but detailed treatments of it can be found in numerous

Table 2.8. *Approximate Values of* $\omega_i \sqrt{m\ell^4/(EI)}$ *for*
$i = 1, 2,$ *and* 3, *for a Clamped–Free Beam Using n Terms of*
a Power Series with a Reduced-Order Equation of Motion

	Mode		
n	1	2	3
1	7.48331	—	—
2	3.84000	57.2822	—
3	3.44050	24.1786	188.677
4	3.52131	20.3280	69.3819
5	3.51698	22.0793	53.2558
6	3.51607	22.1525	61.0295
Exact	3.51602	22.0345	61.6972

graduate level texts on structural dynamics.) In those cases, the Ritz and Galerkin's methods give the same results when used with the same assumed mode functions. As we see here, however, Galerkin's method provides a viable alternative to the Ritz method in cases where the equations of motion are not of the form presented earlier in this chapter.

2.5   Epilogue

In this chapter we have considered the free-vibration analysis and modal representation for flexible structures, along with methods for solving initial-value and forced response problems associated therewith. Moreover, modal approximation methods based on the Ritz and Galerkin methods were introduced. This sets the stage for consideration of aeroelastic problems in Chapters 3 and 4. The static aeroelasticity problem, treated in Chapter 3, results from interaction of structural and aerodynamic loads. These loads are a subset of those involved in dynamic aeroelasticity, which includes inertial effects. One aspect of dynamic aeroelasticity is flutter, which is treated in Chapter 4. It will be seen that both the modal representation and the modal approximation methods apply equally well to both types of problems.

Problems

1. By evaluating the appropriate integrals, prove that each function in the following two sets of functions is orthogonal to all other functions in its set over the interval $0 \leq x \leq \ell$:
   (a) $\sin\left(\frac{n\pi x}{\ell}\right)$ for $n = 1, 2, 3, \ldots$
   (b) $\cos\left(\frac{n\pi x}{\ell}\right)$ for $n = 0, 1, 2, \ldots$

   Use of a table of integrals may be helpful.

2. Considering Eq. (2.59), plot the displacement at time $t = 0$ for a varying number of retained modes, showing that as more modes are kept the shape more closely resembles the initial shape of the string given in Fig. 2.4.

3. Compute the propagation speed of elastic torsional deflections along prismatic, homogeneous, isotropic beams with circular cross sections and made of
   (a) aluminum
   (b) steel

Hint: Compare the governing wave equation with that for the uniform string problem, noting that for beams with a circular cross section $J = I_p$.

Answers: (may vary slightly depending on properties used)
(a) 3140 m/s
(b) 3110 m/s

4. For a uniform string attached between two walls with no external loads, determine the total string deflection $v(x, t)$ for an initial string deflection of zero and an initial transverse velocity distribution given by

$$\frac{\partial v}{\partial t}(x, 0) = V\left[1 - \cos\left(\frac{2\pi x}{\ell}\right)\right].$$

Answer:

$$v(x, t) = -\frac{16V\ell}{\pi^2}\sqrt{\frac{m}{T}}\sum_{n=1,3,\ldots}^{\infty}\frac{1}{n^2(n^2 - 4)}\sin\left(\frac{n\pi x}{\ell}\right)\sin(\omega_n t),$$

where

$$\omega_n = \frac{n\pi}{\ell}\sqrt{\frac{T}{m}}$$

5. Consider a uniform string of length ℓ and mass per unit length m that has been stretched between two walls with tension T. Transverse vibration of the string is restrained at its midpoint by a linear spring with spring constant k. The spring is unstretched when the string is undeflected. Write the generalized equation of motion for the ith mode, giving particular attention to the writing of the generalized force Ξ_i. As a check, derive the equation taking the spring into account through the potential energy instead of through the generalized force.

Answer: Letting

$$\omega_i = \frac{i\pi}{\ell}\sqrt{\frac{T}{m}},$$

one finds that the generalized equations of motion are

$$\ddot{\xi}_i + \omega_i^2\xi_i + \frac{2k}{m\ell}(-1)^{\frac{i-1}{2}}\sum_{j=1,3,\ldots}^{\infty}(-1)^{\frac{j-1}{2}}\xi_j = 0 \qquad (i = 1, 3, \ldots, \infty),$$

$$\ddot{\xi}_i + \omega_i^2\xi_i = 0 \qquad (i = 2, 4, \ldots, \infty).$$

6. Consider a uniform string of length ℓ with mass per unit length m that has been stretched between two walls with tension T. Up until the time $t = 0$ the string is at rest. At time $t = 0$ concentrated loads of magnitude $F_0 \sin \Omega t$ are applied at $x = \ell/3$ and $x = 2\ell/3$ in the positive (up) and negative (down) directions, respectively. In addition, a distributed force

$$F = \overline{F}\left[1 - \sin\left(\frac{3\pi x}{\ell}\right)\right]\cos(\Omega t)$$

is applied to the string. What is the total string displacement $v(x, t)$ for time $t > 0$?

Answer: Letting

$$\omega_n = \frac{n\pi}{\ell}\sqrt{\frac{T}{m}},$$

one finds that

$$v(x,t) = \sum_{n=2,4,\ldots}^{\infty}\left\{C_n\left[\sin(\Omega t) - \frac{\Omega}{\omega_n}\sin(\omega_n t)\right]\sin\left(\frac{n\pi x}{\ell}\right)\right\}$$

$$+ \sum_{n=1,3,\ldots}^{\infty}\left\{D_n\left[\cos(\Omega t) - \cos(\omega_n t)\right]\sin\left(\frac{n\pi x}{\ell}\right)\right\},$$

where

$$C_n = \frac{2F_0}{m\ell\left(\omega_n^2 - \Omega^2\right)}\left[\sin\left(\frac{n\pi}{3}\right) - \sin\left(\frac{2n\pi}{3}\right)\right],$$

$$D_n = \frac{2\overline{F}}{m\left(\omega_n^2 - \Omega^2\right)}\left(\frac{2}{n\pi} - \frac{\delta_{n3}}{2}\right),$$

and where the Kronecker symbol $\delta_{ij} = 1$ for $i = j$ and $\delta_{ij} = 0$ for $i \neq j$.

7. Consider a uniform circular rod of length ℓ, torsional rigidity GJ, and mass moment of inertia per unit length ρJ. The beam is clamped at the end $x = 0$ and it has a concentrated inertia I_c at its other end where $x = \ell$.
   (a) Determine the characteristic equation that can be solved for the torsional natural frequencies for the case in which $I_c = \rho J\ell\zeta$, where ζ is a dimensionless parameter.
   (b) Verify that the characteristic equation obtained in part (a) approaches that obtained in the text for the clamped–free uniform rod in torsion as ζ approaches zero.
   (c) Solve the characteristic equation obtained in part (a) for numerical values of the first four eigenvalues, $\alpha_i\ell$, $i = 1, 2, 3$, and 4, when $\zeta = 1$.
   (d) Solve the characteristic equation obtained in part (a) for the numerical value of the first eigenvalue, $\alpha_1\ell$, when $\zeta = 1, 2, 4$, and 8. Make a plot of the behavior of the lowest natural frequency versus the value of the concentrated inertia. Note that $\alpha_1\ell$ versus ζ is the same thing in terms of dimensionless quantities.

   Answers:
   (a) $\zeta\alpha\ell\tan(\alpha\ell) = 1$
   (c,d) Sample result: $\alpha_1\ell = 0.860334$ for $\zeta = 1$

8. Consider a clamped–free beam undergoing torsion.
   (a) Prove that the free-vibration mode shapes are orthogonal, regardless of whether the beam is uniform.
   (b) Given that the kinetic energy is

$$K = \frac{1}{2}\int_0^\ell \rho I_p\left(\frac{\partial\theta}{\partial t}\right)^2 dx$$

   show that K can be written as

$$K = \frac{1}{2}\sum_{i=1}^{\infty}M_i\dot{\xi}_i^2.$$

   where M_i is the generalized mass of the ith mode and ξ_i is the generalized coordinate for the ith mode.

(c) Given that the potential energy is the internal (i.e., strain) energy, that is,

$$P = \frac{1}{2} \int_0^\ell GJ \left(\frac{\partial \theta}{\partial x} \right)^2 dx,$$

show that P can be written as

$$P = \frac{1}{2} \sum_{i=1}^{\infty} M_i \omega_i^2 \xi_i^2,$$

where ω_i is the natural frequency.

(d) Show that, for a uniform beam with ϕ_i as given in the text, $M_i = \rho I_p \ell / 2$ for all i.

9. Consider a free–free beam undergoing torsion.
   (a) Given the mode shapes found in the text, find an expression for P in terms of GJ, ℓ, and the generalized coordinates.
   (b) Given the mode shapes found in the text, find an expression for K in terms of ρI_p, ℓ, and the time derivatives of the generalized coordinates.
   (c) Substitute your results from parts (a) and (b) into Lagrange's equations and identify the resulting generalized masses.

   Answers:

   (a) $P = \frac{1}{2} GJ \sum_{i=1}^{\infty} \frac{(i\pi)^2}{2\ell} \xi_i^2$

   (b) $K = \frac{1}{2} \rho I_p \ell \left(\dot{\xi}_0^2 + \frac{1}{2} \sum_{i=1}^{\infty} \dot{\xi}_i^2 \right)$

   (c) $M_0 = \rho I_p \ell$; $M_i = \frac{1}{2} \rho I_p \ell$ for $i = 1, 2, \ldots$

10. Consider a clamped–free beam undergoing bending.
    (a) Prove that the free-vibration mode shapes are orthogonal, regardless of whether the beam is uniform.
    (b) Given the kinetic energy as

$$K = \frac{1}{2} \int_0^\ell m \left(\frac{\partial v}{\partial t} \right)^2 dx,$$

show that K can be written as

$$K = \frac{1}{2} \sum_{i=1}^{\infty} M_i \dot{\xi}_i^2,$$

where M_i is the generalized mass of the ith mode, and ξ_i is the generalized coordinate for the ith mode.

(c) Given that the potential energy is the internal (i.e., strain) energy, that is,

$$P = \frac{1}{2} \int_0^\ell EI \left(\frac{\partial^2 v}{\partial x^2} \right)^2 dx,$$

show that P can be written as

$$P = \frac{1}{2} \sum_{i=1}^{\infty} M_i \omega_i^2 \xi_i^2,$$

where ω_i is the natural frequency.

(d) Show that, for a uniform beam and ϕ_i as given in the text, $M_i = m\ell$ for all i.

11. Consider a uniform beam with the boundary conditions shown in Fig. 2.35 undergoing bending vibration.
    (a) Using the relations derived in the text, plot the square of characteristic value, $(\alpha_1 \ell)^2$, which is proportional to the fundamental frequency, versus κ from 0 to 100. Check your results versus those given in Fig. 2.37.
    (b) Plot the fundamental mode shape for values of κ of 0.01, 0.1, 1, 10, and 100. Suggestion: Use Eq. (2.263). You can check your results for $\kappa = 1$ with those given in Fig. 2.36, and your results for $\kappa = 100$ will not differ very much from those of Fig. 2.38 in which $\kappa = 50$.

12. Find the free-vibration frequencies and plot the mode shapes for the first five modes of a beam of length ℓ, having bending stiffness EI and mass per unit length m, that is free at its right end and that has the sliding condition (see Fig. 2.27) at its left end. Normalize the mode shapes to have unit deflection at the free end, and determine the generalized mass for the first five modes.

    Answers: $\omega_0 = 0$, $\omega_1 = 5.59332\sqrt{EI/(m\ell^4)}$, $\omega_2 = 30.2258\sqrt{EI/(m\ell^4)}$, $\omega_3 = 74.6389\sqrt{EI/(m\ell^4)}$, $\omega_4 = 138.791\sqrt{EI/(m\ell^4)}$; $M_0 = m\ell$ and $M_i = m\ell/4$ for $i = 1, 2, \ldots, \infty$. As a sample of the mode shapes, the first elastic mode is plotted in Fig. 2.41.

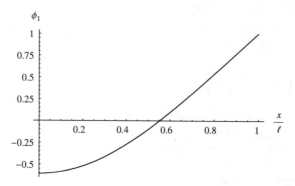

Figure 2.41   First elastic mode shape for sliding–free beam. (Note: the "zeroth" mode is a rigid-body translation mode.)

13. Consider the beam in Problem 12. Add to it a translational spring restraint at the left end, having spring constant $k = \kappa EI/\ell^3$. Find the first three free-vibration frequencies and mode shapes for the cases in which κ takes on values of 0.01, 1, and 100. Plot the mode shapes, normalizing them to have unit deflection at the free end.

    Answers: Sample results: A plot versus κ of $(\alpha_i \ell)^2$ for $i = 1, 2$, and 3 is shown in Fig. 2.42, and the first mode shape for $\kappa = 1$ is shown in Fig. 2.43.

14. Consider a beam that at its left end is clamped and at its right end is pinned with a rigid body attached to it. Let the mass moment of inertia of the rigid body be given by $I_c = \mu m\ell^3$.
    (a) Find the first two free-vibration frequencies for values of μ equal to 0.01, 0.1, 1, 10, and 100. Comment on the variation of the natural frequencies versus μ.
    (b) Choose any normalization that is convenient, and plot the first mode shape for these same values of μ. Comment on the variation of the mode shapes versus μ.

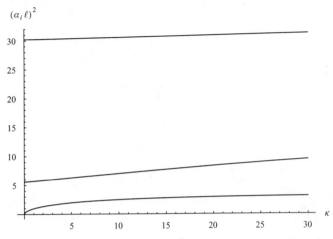

Figure 2.42   Variation versus κ of $(\alpha_i \ell)^2$ for $i = 1, 2$, and 3, for a beam that is free on its right end and has a sliding boundary condition spring-restrained in translation on its left end.

Answers:
(a) Sample result: $\omega_1 = 1.99048 \sqrt{EI/(m\ell^4)}$ for $\mu = 1$.
(b) Sample result: The first mode shape for $\mu = 1$ is shown in Fig. 2.44.

15. Consider a uniform cantilever beam of length ℓ, bending rigidity EI, and mass per unit length m. Until time $t = 0$ the beam is undeflected and at rest. At time $t = 0$ a transverse concentrated load of magnitude $F \cos(\Omega t)$ is applied at $x = \ell$.
    (a) Write the generalized equations of motion.
    (b) Determine the total beam displacement $v(x, t)$ for time $t > 0$.
    (c) For the case when $\Omega = 0$, determine the tip displacement of the beam. Ignoring those terms that are time dependent (these would die out in a real beam because of dissipation), plot the tip displacement versus the number of mode shapes retained in the solution up to five modes. Show the static tip deflection from elementary beam theory on the plot. (This part of the problem illustrates how the modal representation can be applied to static response problems.)

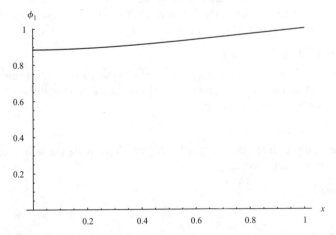

Figure 2.43   First mode shape for a beam that is free on its right end and has a sliding boundary condition spring-restrained in translation on its left end with $\kappa = 1$.

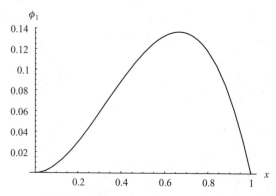

Figure 2.44   First mode shape for a beam that is clamped on its left end and pinned with a rigid body attached on its right end with $\mu = 1$.

Answers:

(a) The ith equation is

$$\ddot{\xi}_i + \omega_i^2 \xi_i = 2(-1)^{i+1} \frac{F}{m\ell} \cos(\Omega t).$$

(b) With $\phi_i(x)$ given by Eq. (2.251), $\omega_i = (\alpha_i\ell)^2 \sqrt{EI/(m\ell^4)}$, and $\alpha_i\ell$ as given in Table 2.1, one finds that

$$v(x,t) = \frac{2F}{m\ell} \sum_{i=1}^{\infty} \frac{(-1)^{i+1}}{\omega_i^2 - \Omega^2} \left[\cos(\Omega t) - \cos(\omega_i t)\right] \phi_i(x).$$

(c) The result converges within engineering accuracy to $F\ell^3/(3EI)$ using only a few terms.

16. Consider a free–free beam with bending stiffness EI, mass per unit length m, and length ℓ. Write the generalized equations of motion for a system that consists of the beam plus identical rigid bodies attached to the ends, where each body has a moment of inertia I_c and mass m_c. Use as assumed modes those of the exact solution of the free–free beam without the attached bodies, obtained in the text. Note the terms that provide inertial coupling.

17. Consider a pinned–free beam with the rotation about the hinge restrained by a light spring of modulus $\kappa EI/\ell$. Use a rigid-body rotation plus the set of clamped–free modes as the assumed modes of the Ritz method. Compare the results for the first two modes using a varying number of terms for $\kappa = 1$, 10, and 100.

    Answer: See Tables 2.9 and 2.10.

18. Repeat Problem 17 using a set of polynomial admissible functions. Use one rigid-body mode (x) and a varying number of polynomials that satisfy the boundary conditions of a clamped–free beam.

    Answers: See Tables 2.11 and 2.12.

19. Consider a clamped–free beam of length ℓ for which the mass per unit length and bending stiffness vary according to

$$m = m_0 \left(1 - \frac{x}{\ell} + \mu\frac{x}{\ell}\right),$$

$$EI = EI_0 \left(1 - \frac{x}{\ell} + \kappa\frac{x}{\ell}\right).$$

Table 2.9. *Approximate Values of $\omega_1\sqrt{m\ell^4/(EI)}$ for Pinned–Free Beam Having a Root Rotational Spring with Spring Constant of $\kappa EI/\ell$ Using One Rigid-Body Mode (x) and $n - 1$ Clamped–Free Modes of Section 2.3.4, Eq. (2.251)*

	κ		
n	1	10	100
1	1.73205	5.47723	17.3205
2	1.55736	2.96790	3.44766
3	1.55730	2.96784	3.44766
4	1.55730	2.96784	3.44766
5	1.55730	2.96784	3.44766
Exact	1.55730	2.96784	3.44766

Table 2.10. *Approximate Values of $\omega_2\sqrt{m\ell^4/(EI)}$ for Pinned–Free Beam Having a Root Rotational Spring with Spring Constant of $\kappa EI/\ell$ Using One Rigid-Body Mode (x) and $n - 1$ Clamped–Free Modes of Section 2.3.4, Eq. (2.251)*

	κ		
n	1	10	100
2	22.8402	37.9002	103.173
3	16.2664	19.3632	21.6202
4	16.2512	19.3563	21.6200
5	16.2502	19.3559	21.6200
Exact	16.2501	19.3558	21.6200

Table 2.11. *Approximate Values of $\omega_1\sqrt{m\ell^4/(EI)}$ for Pinned–Free Beam Having a Root Rotational Spring with Spring Constant of $\kappa EI/\ell$ Using One Rigid-Body Mode (x) and $n - 1$ Polynomials That Satisfy Clamped–Free Beam Boundary Conditions*

	κ		
n	1	10	100
1	1.73205	5.47723	17.3205
2	1.55802	2.97497	3.46064
3	1.55730	2.96784	3.44768
4	1.55730	2.96784	3.44766
Exact	1.55730	2.96784	3.44766

Using the comparison functions in Eq. (2.310), apply the Ritz method to determine approximate values for the first three natural frequencies, varying the number of terms from one to five. Let $\mu = \kappa = 1/2$.

Answer: See Table 2.13.

Table 2.12. *Approximate Values of $\omega_2 \sqrt{m\ell^4/(EI)}$ for Pinned–Free Beam Having a Root Rotational Spring with Spring Constant of $\kappa EI/\ell$ Using One Rigid-Body Mode (x) and $n - 1$ Polynomials That Satisfy Clamped–Free Beam Boundary Conditions*

		κ	
n	1	10	100
2	24.8200	41.1049	111.743
3	16.4047	19.7070	22.2338
4	16.2508	19.3565	21.6208
Exact	16.2501	19.3558	21.6200

Table 2.13. *Approximate Values of $\omega_i \sqrt{m_0\ell^4/(EI_0)}$ for a Tapered, Clamped–Free Beam Based on the Ritz Method with n Polynomials That Satisfy Clamped–Free Beam Boundary Conditions*

n	$\omega_1 \sqrt{\frac{m_0\ell^4}{EI_0}}$	$\omega_2 \sqrt{\frac{m_0\ell^4}{EI_0}}$	$\omega_3 \sqrt{\frac{m_0\ell^4}{EI_0}}$
1	4.36731	—	—
2	4.31571	24.7653	—
3	4.31517	23.5267	69.8711
4	4.31517	23.5199	63.2441
5	4.31517	23.5193	63.2415
Exact	4.31517	23.5193	63.1992

20. Rework Problem 19 using the Ritz method and the set of polynomial admissible functions $(x/\ell)^{i+1}$, $i = 1, 2, \ldots, n$.

Answer: See Table 2.14.

Table 2.14. *Approximate Values of $\omega_i \sqrt{m_0\ell^4/(EI_0)}$ for a Tapered, Clamped–Free Beam Based on the Ritz Method with n Terms of the Form $(x/\ell)^{i+1}$, $i = 1, 2, \ldots, n$*

n	$\omega_1 \sqrt{\frac{m_0\ell^4}{EI_0}}$	$\omega_2 \sqrt{\frac{m_0\ell^4}{EI_0}}$	$\omega_3 \sqrt{\frac{m_0\ell^4}{EI_0}}$
1	5.07093	—	—
2	4.31883	33.8182	—
3	4.31732	23.6645	110.529
4	4.31523	23.6640	64.8395
5	4.31517	23.5226	64.7821
Exact	4.31517	23.5193	63.1992

21. Rework Problem 19 using Galerkin's method and Eq. (2.315) with $f = 0$ as the equation of motion and the set of polynomial admissible functions $(x/\ell)^{i+1}$, $i = 1, 2, \ldots, n$.

Answer: See Table 2.15.

Table 2.15. *Approximate Values of* $\omega_i \sqrt{m_0 \ell^4/(E I_0)}$ *for a Tapered, Clamped–Free Beam Based on the Galerkin Method Applied to Eq. (2.315) with n Terms of the Form* $(x/\ell)^{i+1}, i = 1, 2, \ldots, n$

n	$\omega_1 \sqrt{\frac{m_0 \ell^4}{E I_0}}$	$\omega_2 \sqrt{\frac{m_0 \ell^4}{E I_0}}$	$\omega_3 \sqrt{\frac{m_0 \ell^4}{E I_0}}$
1	7.88811	—	—
2	4.45385	54.5221	—
3	4.19410	24.3254	175.623
4	4.33744	21.4784	67.1265
5	4.31379	23.8535	53.6214
Exact	4.31517	23.5193	63.1992

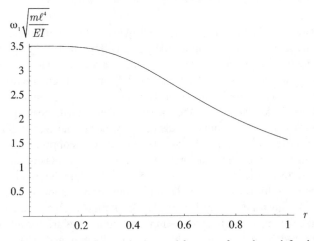

Figure 2.45   Approximate fundamental frequency for a clamped–free beam with a particle of mass $m\ell$ attached at $x = r\ell$.

22. Consider a clamped–free beam to which is attached at spanwise location $x = \ell r$ a particle of mass $\mu m \ell$. Using a two-term Ritz approximation based on the functions in Eq. (2.311), plot the approximate value for the fundamental natural frequency as a function of r for $\mu = 1$.

Answer: See Fig. 2.45.

CHAPTER 3

Static Aeroelasticity

I discovered that with increasing load, the angle of incidence at the wing tips increased perceptibly. It suddenly dawned on me that this increasing angle of incidence was the cause of the wing's collapse, as logically the load resulting from the air pressure in a steep dive would increase faster at the wing tips than at the middle. The resulting torsion caused the wings to collapse under the strain of combat maneuvers.

A. H. G. Fokker in *The Flying Dutchman* (Henry Holt and Company, 1931)

The field of static aeroelasticity is the study of flight vehicle phenomena associated with the interaction of aerodynamic loading induced by steady flow and the resulting elastic deformation of the lifting surface structure. These phenomena are characterized as being insensitive to the rates and accelerations of the structural deflections. There are two classes of design problems that are encountered in this area. The first and most common to all flight vehicles is the effect of elastic deformation on the airloads associated with normal operating conditions. These effects can have a profound influence on the performance, handling qualities, flight stability, and structural load distribution. The second class of problems involves the potential for static instability of the structure that will result in a catastrophic failure. This instability is often termed "divergence" and can impose a limit on the flight envelope.

The material presented in this chapter provides an introduction to some of these static aeroelastic phenomena. To illustrate the physical mechanics of these problems and maintain a low level of mathematical complexity, relatively simple configurations are considered. The first items treated are rigid aerodynamic models that are elastically mounted in a wind tunnel test section. Such elastic mounting is characteristic of most load measurement systems. The second aeroelastic configuration to be treated is a uniform elastic lifting surface of finite span. Its static aeroelastic properties are quite similar to most lifting surfaces on conventional flight vehicles.

3.1   Wind Tunnel Models

In this section we consider three types of mounting for wind tunnel models: wall-mounted, sting-mounted, and strut-mounted. Expressions for the aeroelastic pitch deflections are developed for these simple models that, in turn, lead us to a cursory understanding of the divergence instability. Finally, we will return to the wall-mounted model briefly in this section to consider the qualitatively different phenomenon of aileron reversal.

3.1.1   *Wall-Mounted Model*

Consider a rigid, spanwise-uniform model of a wing that is mounted to the side walls of a wind tunnel in such a way as to allow the wing to pitch about the support axis, as illustrated in Fig. 3.1. The support is flexible in torsion, which means that it restricts the pitch rotation of the wing in the same way as a rotational spring would. We denote the rotational stiffness of the support by k; see Fig. 3.2. If we assume the body to be pivoted

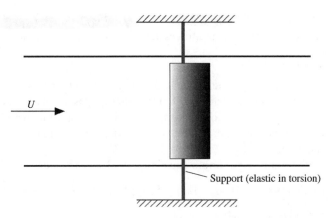

Figure 3.1   Planform view of a wind tunnel model on a torsionally elastic support.

about its support O, located at a distance x_O from the leading edge, moment equilibrium requires that the sum of all moments about O must equal zero. In anticipation of using linear aerodynamics, we assume the angle of attack, α, to be a small angle, such that $\sin(\alpha) = \alpha$ and $\cos(\alpha) = 1$. Thus,

$$M_{ac} + L(x_O - x_{ac}) - W(x_O - x_{cg}) - k\theta = 0. \qquad (3.1)$$

If the support were rigid, the angle of attack would be α_r, positive nose-up. The elastic part of the pitch angle is denoted by θ, which is also positive nose-up. The wing angle of attack is then $\alpha \equiv \alpha_r + \theta$. For linear aerodynamics, the lift for a rigid support is simply

$$L_{rigid} = qSC_{L_\alpha}\alpha_r, \qquad (3.2)$$

whereas the lift for an elastic support is

$$L = qSC_{L_\alpha}(\alpha_r + \theta), \qquad (3.3)$$

where $q = \frac{1}{2}\rho_\infty U^2$ is the freestream dynamic pressure (i.e., in the far field – often denoted by q_∞), U is the freestream air speed, ρ_∞ is the freestream air density, S is the planform area,

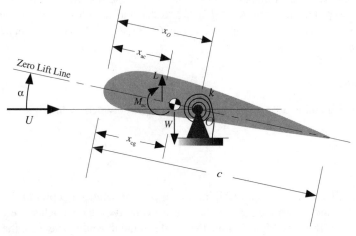

Figure 3.2   Airfoil for wind tunnel model.

and C_{L_α} is the wing lift-curve slope. Note that $L \neq L_{\text{rigid}}$; and, for positive θ, $L > L_{\text{rigid}}$. We can express the moment of aerodynamic forces about the aerodynamic center as

$$M_{\text{ac}} = qScC_{M\text{ac}}. \tag{3.4}$$

If the angle of attack is small, $C_{M\text{ac}}$ can be regarded as a constant. It should be noted here that linear aerodynamics implies that the lift-curve slope C_{L_α} is a constant. A further simplification may be that $C_{L_\alpha} = 2\pi$ in accordance with two-dimensional thin-airfoil theory. If experimental data or results from computational fluid dynamics provide an alternative value, then it should be used.

Using Eqs. (3.3) and (3.4), the equilibrium equation, Eq. (3.1), can be expanded as

$$qScC_{M\text{ac}} + qSC_{L_\alpha}(\alpha_{\text{r}} + \theta)(x_O - x_{\text{ac}}) - W(x_O - x_{\text{cg}}) = k\theta. \tag{3.5}$$

Solving Eq. (3.5) for the elastic deflection, one obtains

$$\theta = \frac{qScC_{M\text{ac}} + qSC_{L_\alpha}\alpha_{\text{r}}(x_O - x_{\text{ac}}) - W(x_O - x_{\text{cg}})}{k - qSC_{L_\alpha}(x_O - x_{\text{ac}})}. \tag{3.6}$$

When α_{r} and q are specified the total lift can be determined.

When the support point O is aft of the aerodynamic center, so that $x_O > x_{\text{ac}}$, the denominator can vanish, which implies that θ blows up. This behavior is a static aeroelastic instability called "divergence." The dynamic pressure at which divergence occurs for this case is

$$q_D = \frac{k}{SC_{L_\alpha}(x_O - x_{\text{ac}})}. \tag{3.7}$$

From this, the air speed at which divergence occurs can be found as

$$U_D = \sqrt{\frac{2k}{\rho_\infty SC_{L_\alpha}(x_O - x_{\text{ac}})}}. \tag{3.8}$$

It is evident that when the aerodynamic center is coincident with the pivot, so that $x_O = x_{\text{ac}}$, the divergence dynamic pressure becomes infinite. Also, when the aerodynamic center is aft of the pivot, so that $x_O < x_{\text{ac}}$, the divergence dynamic pressure becomes negative. In either case divergence is impossible.

To pursue the character of this instability a bit further, consider the case of a symmetric airfoil ($C_{M\text{ac}} = 0$). Furthermore, let $x_O = x_{\text{cg}}$ so that the weight term drops out of the equation for θ. From Eq. (3.7) we can let $k = q_D SC_{L_\alpha}(x_O - x_{\text{ac}})$, and so θ can be written simply as

$$\theta = \frac{\alpha_{\text{r}}}{\frac{q_D}{q} - 1}. \tag{3.9}$$

The lift is proportional to $\alpha_{\text{r}} + \theta$. Thus, the change in lift divided by the rigid lift is given by

$$\frac{\Delta L}{L_{\text{rigid}}} = \frac{\theta}{\alpha_{\text{r}}} = \frac{\frac{q}{q_D}}{1 - \frac{q}{q_D}}. \tag{3.10}$$

Both θ and $\Delta L/L_{\text{rigid}}$ clearly approach infinity as $q \to q_D$. Indeed, a plot of the latter is given in Fig. 3.3 and shows the large change in lift caused by the aeroelastic effect. The lift evidently starts from its "rigid" value, that is, the value it would have were the support rigid, and increases to infinity as $q \to q_D$. However, keep in mind that there are limitations on the validity of both expressions. Namely, the lift will not continue to increase as stall is

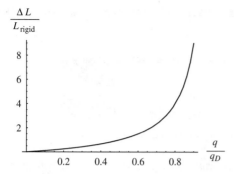

Figure 3.3   Relative change in lift caused by aeroelastic effect.

encountered. Moreover, the structure will not tolerate infinite deformation, and failure will take place at some finite value of θ.

When the system parameters are within the bounds of validity for linear theory, another fascinating feature of this problem emerges. One can invert the expression for θ to obtain

$$\frac{1}{\theta} = \frac{q_D}{\alpha_r}\left(\frac{1}{q} - \frac{1}{q_D}\right),\tag{3.11}$$

making it evident that $1/\theta$ is proportional to $1/q$ (see Fig. 3.4). Therefore, for a model of this type only two data points are needed to extrapolate the line down and to the left until it

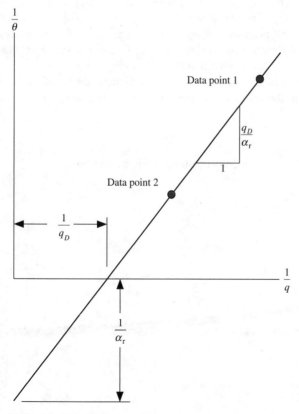

Figure 3.4   Plot of $1/\theta$ versus $1/q$.

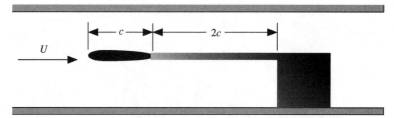

Figure 3.5 Schematic of a sting-mounted wind tunnel model.

intercepts the $1/q$ axis at a distance $1/q_D$ from the origin. As can be seen from the figure, the slope of this line can also be used to estimate q_D. The form of this plot is of great practical value because estimates of q_D can be extrapolated from data taken at speeds far below the divergence speed. This means that q_D can be estimated even when the values of the model parameters are not precisely known, thus circumventing the need to risk destruction of the model by testing all the way up to the divergence boundary.

3.1.2   *Sting-Mounted Model*

A second configuration of potential interest is a rigid model mounted on an elastic sting. A simplified version of this kind of model is shown in Figs. 3.5–3.7, where the sting is modeled as a uniform, elastic, cantilevered beam with bending stiffness EI and length $2c$. The model is mounted in such a way as to have angle of attack of α_r when the beam is undeformed. Thus, as before $\alpha = \alpha_r + \theta$, where θ is the nose-up rotation of the wing resulting from bending of the sting, as shown in Fig. 3.6. Also in Fig. 3.6 we denote the tip deflection of the cantilever beam as δ, although we do not need it for this analysis. One should note the equal and opposite directions on the force F_0 and moment M_0 at the trailing edge of the wing in Fig. 3.7 versus at the tip of the sting in Fig. 3.6.

From superposition one can deduce the total bending slope at the tip of the sting as the sum of contributions from the tip force F_0 and tip moment M_0, denoted by θ_F and θ_M, respectively, so that

$$\theta = \theta_F + \theta_M. \tag{3.12}$$

From elementary beam theory these constituent parts can be written as

$$\theta_F = \frac{F_0(2c)^2}{2EI},$$

$$\theta_M = \frac{M_0(2c)}{EI}, \tag{3.13}$$

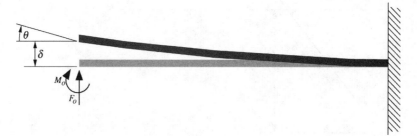

Figure 3.6 Detailed view of the cantilevered beam.

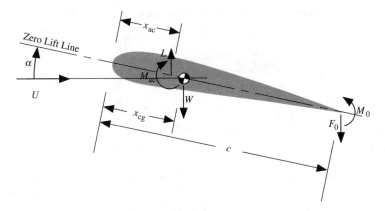

Figure 3.7   Detailed view of the sting-mounted wing.

and so

$$F_0 = \frac{2EI\,\theta_F}{(2c)^2},$$
$$M_0 = \frac{EI\,\theta_M}{2c}. \tag{3.14}$$

Two static aeroelastic equilibrium equations can now be written for the determination of θ_F and θ_M. Using Eqs. (3.3) and (3.4) for the lift and pitching moment, the force equilibrium equation can be written as

$$q S C_{L_\alpha}(\alpha_r + \theta_F + \theta_M) - W - F_0 = 0, \tag{3.15}$$

and the sum of moments about the trailing edge yields

$$q S c C_{Mac} + q S C_{L_\alpha}(\alpha_r + \theta_F + \theta_M)(c - x_{ac}) - W(c - x_{cg}) - M_0 = 0. \tag{3.16}$$

Substitution of Eqs. (3.14) into Eqs. (3.15) and (3.16), simultaneous solution for θ_F and θ_M, and use of Eq. (3.12) yields

$$\theta = \frac{\alpha_r\left(2 - \frac{x_{ac}}{c}\right) + \frac{C_{Mac}}{C_{L_\alpha}} - \frac{W}{qSC_{L_\alpha}}\left(2 - \frac{x_{cg}}{c}\right)}{\frac{EI}{2qSc^2C_{L_\alpha}} - \left(2 - \frac{x_{ac}}{c}\right)}. \tag{3.17}$$

Here again the condition for divergence can be obtained by setting the denominator to zero, so that

$$q_D = \frac{EI}{2\left(2 - \frac{x_{ac}}{c}\right)Sc^2C_{L_\alpha}}. \tag{3.18}$$

However, unlike the previous example, one cannot make the divergence dynamic pressure infinite or negative (thereby making divergence mathematically impossible) by choice of configuration parameters because $x_{ac}/c \leq 1$. For a given wing configuration, one is left only with the possibility of increasing the sting bending stiffness to make the divergence dynamic pressure larger.

3.1.3   *Strut-Mounted Model*

A third configuration of a wind tunnel mount is a strut system as idealized in Figs. 3.8 and 3.9. The two linearly elastic struts have the same extensional stiffness, k, and

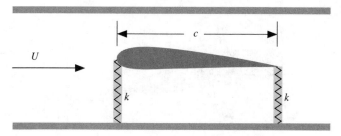

Figure 3.8 Schematic of strut-supported wind tunnel model.

are mounted at the leading and trailing edges of the wing. The model is mounted in such a way as to have angle of attack of α_r when the springs are both undeformed. Thus, as before, the angle of attack is $\alpha = \alpha_r + \theta$. As illustrated in Fig. 3.9, the elastic pitch angle, θ, can be related to the extension of the two struts as

$$\theta = \frac{\delta_1 - \delta_2}{c}. \tag{3.19}$$

The sum of the forces in the vertical direction shows that

$$L - W - k(\delta_1 + \delta_2) = 0. \tag{3.20}$$

The sum of the moments about the trailing edge yields

$$M_{ac} + L(c - x_{ac}) - W(c - x_{cg}) - kc\delta_1 = 0. \tag{3.21}$$

Again, using Eqs. (3.3) and (3.4) for the lift and pitching moment, the simultaneous solution of the force and moment equations yields

$$\theta = \frac{\alpha_r \left(1 - 2\frac{x_{ac}}{c}\right) + 2\frac{C_{Mac}}{C_{L_\alpha}} - \frac{W}{qSC_{L_\alpha}}\left(1 - 2\frac{x_{cg}}{c}\right)}{\frac{kc}{qSC_{L_\alpha}} - \left(1 - 2\frac{x_{ac}}{c}\right)}. \tag{3.22}$$

As usual, the divergence condition is indicated by the vanishing of the denominator, so that

$$q_D = \frac{kc}{SC_{L_\alpha}\left(1 - 2\frac{x_{ac}}{c}\right)}. \tag{3.23}$$

It is evident for this problem as specified that when the aerodynamic center is in front of the mid-chord (as it is in subsonic flow), the divergence condition cannot be eliminated. However, divergence can be eliminated if the leading-edge spring stiffness is increased relative to that of the trailing-edge spring. This is left as an exercise for the reader; see Problem 5.

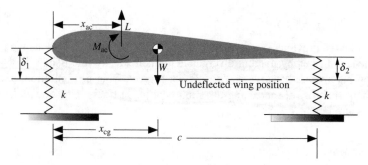

Figure 3.9 Cross section of strut-supported wind tunnel model.

3.1.4   Wall-Mounted Model for Application to Aileron Reversal

Before putting aside the wind tunnel type models dealt with so far in this chapter, we here consider the problem of aileron reversal. It is known that wing torsional flexibility causes certain primary flight control devices, such as ailerons, to function in a manner that is completely at odds with their intended purpose. The primary danger posed by the loss of control effectiveness is that the pilot cannot control the aircraft in the usual way. There are additional concerns for aircraft, the missions of which depend on their being highly maneuverable. For example, when control effectiveness is lost, the pilot may not be able to count on the aircraft's ability to execute evasive maneuvers. It is this loss in control effectiveness and eventual reversal that is the focus of this section.

To this end, consider the airfoil section of a flapped two-dimensional wing, shown in Fig. 3.10. Similar to the model discussed in Section 3.1.1, the wing is pivoted and restrained by a rotational spring with spring constant k. The main differences are that (1) a trailing-edge flap is added such that the flap angle β can be arbitrarily set by the flight control system; and (2) we need not consider gravity to illustrate this phenomenon, so the weight is not shown in the figure. Moment equilibrium for this system about the pivot requires that

$$M_{\mathrm{ac}} + eL = k\theta. \tag{3.24}$$

The lift and pitching moment for a two-dimensional wing can be written as before, namely,

$$\begin{aligned} L &= qSC_L, \\ M_{\mathrm{ac}} &= qcSC_{M\mathrm{ac}}. \end{aligned} \tag{3.25}$$

When $\beta \neq 0$, the effective camber of the airfoil changes, inducing changes in both lift and pitching moment. For a linear theory, both α and β should be small angles, so that

$$\begin{aligned} C_L &= C_{L_\alpha}\alpha + C_{L_\beta}\beta, \\ C_{M\mathrm{ac}} &= C_{M_0} + C_{M_\beta}\beta. \end{aligned} \tag{3.26}$$

Note that $C_{M_\beta} < 0$. For convenience, we assume a symmetric airfoil ($C_{M_0} = 0$).

We now substitute Eqs. (3.25) into the moment equilibrium equation, Eq. (3.24), making use of Eqs. (3.26), and determine θ to be

$$\theta = \frac{qS\left[eC_{L_\alpha}\alpha_{\mathrm{r}} + \left(eC_{L_\beta} + cC_{M_\beta}\right)\beta\right]}{k - eqSC_{L_\alpha}}. \tag{3.27}$$

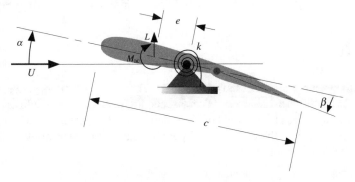

Figure 3.10   Schematic of the airfoil section of a flapped two-dimensional wing in a wind tunnel.

We see that, because of the torsional flexibility (represented here by the rotational spring), θ is a function of β. Substituting Eq. (3.27) back into Eqs. (3.25a) and (3.26a), one obtains an expression for the aeroelastic lift,

$$L = \frac{qS\left[C_{L_\alpha}\alpha_{\mathrm{r}} + C_{L_\beta}\left(1 + \frac{cqSC_{L_\alpha}C_{M_\beta}}{kC_{L_\beta}}\right)\beta\right]}{1 - \frac{eqSC_{L_\alpha}}{k}}. \tag{3.28}$$

It is evident from the two terms in the coefficient of β in this expression that lift is a function of β in two counteracting ways. Ignoring the effect of the denominator for the time being, one sees that the first term in the numerator that multiplies β is purely aerodynamic and leads to an *increase* in lift with β, because of a change in the effective camber. The second term is aeroelastic. Recalling that $C_{M_\beta} < 0$, one sees that as β is increased, the effective change in the camber also induces a nose-down pitching moment that, because the wing is torsionally flexible, tends to decrease θ and in turn *decrease* lift. At low speed, the purely aerodynamic increase in lift overpowers the aeroelastic tendency to decrease the lift, so that the lift indeed increases with β (and the aileron works as advertised). However, as dynamic pressure increases, the aeroelastic effect becomes stronger; and there is a point at which the net rate of change of lift with respect to β vanishes so that

$$\frac{\partial L}{\partial \beta} = 0 = \frac{qSC_{L_\beta}\left(1 + \frac{cqSC_{L_\alpha}C_{M_\beta}}{kC_{L_\beta}}\right)}{1 - \frac{eqSC_{L_\alpha}}{k}}. \tag{3.29}$$

Thus, one finds that the dynamic pressure at which the reversal occurs is

$$q_R = -\frac{kC_{L_\beta}}{cSC_{L_\alpha}C_{M_\beta}}. \tag{3.30}$$

Notice that since $C_{M_\beta} < 0$, $q_R > 0$. Obviously, a stiffer k gives a higher reversal speed, and a torsionally rigid wing will not undergo reversal. For dynamic pressures above q_R (but still below the divergence dynamic pressure), a positive β will actually decrease the lift.

Now let us consider the effect of both numerator and denominator. As before, the denominator in L (and θ) can vanish, resulting in divergence. The dynamic pressure at which divergence occurs can be found by setting the denominator of Eq. (3.28) equal to zero, which yields

$$q_D = \frac{k}{eSC_{L_\alpha}}. \tag{3.31}$$

Equations (3.30) and (3.31) can be used to simplify the expression for the lift in Eq. (3.28) to obtain

$$L = \frac{qS\left[C_{L_\alpha}\alpha_{\mathrm{r}} + C_{L_\beta}\left(1 - \frac{q}{q_R}\right)\beta\right]}{1 - \frac{q}{q_D}}. \tag{3.32}$$

It is clear from this expression that the coefficient of β can be positive, negative, or zero. Thus, a positive β could increase the lift, decrease the lift, or not change the lift at all. The aileron's lift efficiency, η, can be thought of as the aeroelastic (i.e., actual) change in lift per unit change in β divided by the change in lift per unit change in β that would result were the wing not flexible in torsion; that is,

$$\eta = \frac{\text{change in lift per unit change in } \beta \text{ for elastic wing}}{\text{change in lift per unit change in } \beta \text{ for rigid wing}}.$$

Using the above, one can easily find that

$$\eta = \frac{1 - \frac{q}{q_R}}{1 - \frac{q}{q_D}}, \tag{3.33}$$

which implies that the wing will remain divergence-free and control efficiency will not be lost as long as $q < q_D \leq q_R$. Obviously, if the wing were rigid, both q_D and q_R become infinite and $\eta = 1$.

Thinking unconventionally for the moment, let us allow the possibility of $q_R \ll q_D$. This will result in aileron reversal at a low speed, of course. Although the aileron will now work the opposite of the usual way at most of the operational speeds of the aircraft, this type of design should not be ruled out on these grounds alone. Active flight-control systems can certainly compensate for this. Moreover, one can get considerably more (negative) lift for positive β in this unusual regime than positive lift for positive β in the more conventional setting. This may have important implications for development of highly maneuverable aircraft. Exactly what other potential advantages and disadvantages exist from following this strategy, particularly in this era of composite materials, smart structures, and active controls, is not presently known and is the subject of current research.

3.2   Uniform Lifting Surface

Up to now, our aeroelastic analyses have focused on rigid wings with a flexible support. These idealized configurations do give some insight into the aeroelastic stability and response, but practical analyses must take flexibility of the lifting surface into account. That being the case, in this section we address flexible wings, albeit with simplified structural representation.

Consider an unswept uniform elastic lifting surface as illustrated in Figs. 3.11 and 3.12. The lifting surface is modeled as a beam and is presumed to be built in at its root ($y = 0$, to represent attachment to a wind tunnel wall or a fuselage) and free at its tip ($y = \ell$). The y axis corresponds to the elastic axis, which may be defined as the line of effective shear centers, assumed here to be straight. For isotropic beams, a transverse force applied at any point along this axis will result in bending with no elastic torsional rotation about the axis. This axis is also the axis of twist in response to a pure twisting moment applied to the wing. Because the primary concern here is the determination of the airload distributions, the only elastic deformation that will influence these loads is rotation due to twist about the elastic axis.

3.2.1   *Equilibrium Equation*

The total applied moment (per unit span) about this axis will be denoted as $M'(y)$, which is positive leading-edge up and given by

$$M' = M'_{\text{ac}} + eL' - Nmgd, \tag{3.34}$$

where L' and M'_{ac} are the distributed spanwise lift and pitching moment (i.e., the lift and pitching moment per unit length), mg is the spanwise weight distribution (i.e., the weight per unit length), and N is the "normal load factor" for the case in which the wing is level (i.e., the z axis is directed vertically upward). Thus, N can be written as

$$N = \frac{L}{W} = 1 + \frac{A_z}{g}, \tag{3.35}$$

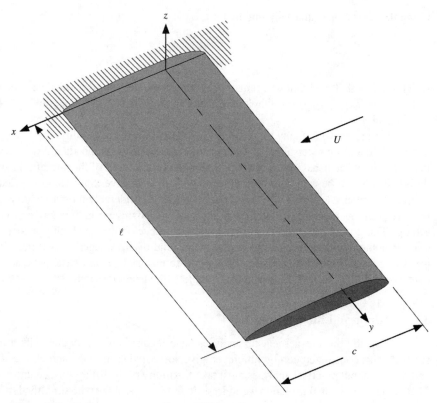

Figure 3.11 Uniform unswept cantilevered lifting surface.

where A_z is the z component of the wing's inertial acceleration, W is the total weight of the aircraft, and L is the total lift. The distributed aerodynamic loads can be written in coefficient form as

$$L' = qcc_\ell,$$
$$M'_{ac} = qc^2 c_{mac}, \tag{3.36}$$

where the freestream dynamic pressure, q, is

$$q = \frac{1}{2}\rho_\infty U^2. \tag{3.37}$$

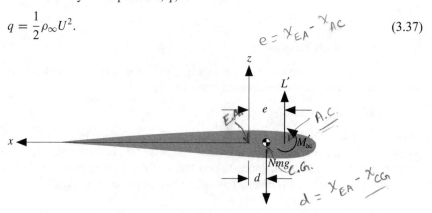

Figure 3.12 Cross section of spanwise uniform lifting surface.

Now, a static equilibrium equation for the elastic torsional rotation, θ, about the elastic axis can be obtained from the fundamental torsional relation

$$T = GJ\frac{d\theta}{dy},\tag{3.38}$$

where GJ is the effective torsional stiffness and T is the twisting moment about the elastic axis. One can obtain an equilibrium equation by equating the rate of change of twisting moment to the negative of the applied torque distribution so that

$$\frac{dT}{dy} = \frac{d}{dy}\left(GJ\frac{d\theta}{dy}\right) = -M'.\tag{3.39}$$

Recognizing that uniformity implies GJ is constant over the length, substituting Eqs. (3.36) into Eq. (3.34) to obtain the applied torque, and finally substituting the applied torque and Eq. (3.38) for the internal torque into the equilibrium equation, Eq. (3.39), one obtains

$$GJ\frac{d^2\theta}{dy^2} = -qc^2 c_{mac} - eqcc_\ell + Nmgd.\tag{3.40}$$

The sectional lift coefficient can be related to the angle of attack by an appropriate aerodynamic theory as some function $c_\ell(\alpha)$, where the functional relation generally involves integration over the planform. To simplify the calculation, the wing can be broken up into spanwise segments of infinitesimal length, where the local lift can be estimated from two-dimensional theory. This theory, commonly known as strip theory, frequently makes use of table look-up for efficient calculation. Here we will use an even simpler, linear form in which the lift-curve slope is assumed to be a constant along the span. Thus,

$$c_\ell(\alpha) = \frac{dc_\ell}{d\alpha}\alpha(y) = a\alpha(y),\tag{3.41}$$

where the constant sectional lift-curve slope is denoted by a.

The angle of attack will be represented by two components. The first is a rigid contribution, α_r, from a "rigid" rotation of the surface (plus any built-in twist, although none is assumed to exist here). The second component is the elastic torsional rotation θ. Hence,

$$\alpha(y) = \alpha_r + \theta(y).\tag{3.42}$$

Associated with each angle of attack contribution is a component of sectional lift coefficient given by strip theory as

$$c_\ell(y) = a[\alpha_r + \theta(y)].\tag{3.43}$$

This aerodynamic representation can be substituted into the equilibrium equation to yield its final form as

$$\frac{d^2\theta}{dy^2} + \frac{qcae}{GJ}\theta = -\frac{1}{GJ}(qc^2 c_{mac} + qcae\alpha_r - Nmgd).\tag{3.44}$$

Finally, a complete description of this equilibrium condition requires specification of the boundary conditions. Since the surface is built-in at the root and free at the tip, these conditions can be written as

$$\begin{aligned} y = 0: &\qquad \theta = 0 &&\text{(zero deflection)},\\[4pt] y = \ell: &\qquad \frac{d\theta}{dy} = 0 &&\text{(zero twisting moment)}. \end{aligned}\tag{3.45}$$

Obviously, these boundary conditions are only valid for the clamped–free condition. The boundary conditions for other end conditions for beams in torsion are given in Chapter 2, Section 2.2.2.

3.2.2 Torsional Divergence

If it is presumed that the configuration parameters of the above uniform wing are known, then it should be possible to solve Eq. (3.44) to determine the resulting twist distribution and associated airload. To simplify the notation let

$$\lambda^2 \equiv \frac{qcae}{GJ},$$

$$\lambda^2 \overline{\alpha}_{\mathrm{r}} \equiv \frac{1}{GJ}(qc^2 c_{\mathrm{mac}} - Nmgd), \tag{3.46}$$

so that

$$\overline{\alpha}_{\mathrm{r}} = \frac{cc_{\mathrm{mac}}}{ae} - \frac{Nmgd}{qcae}. \tag{3.47}$$

Note that λ^2 and $\overline{\alpha}_{\mathrm{r}}$ are independent of y since the wing is assumed to be uniform. The static aeroelastic equilibrium equation can now be written as

$$\frac{d^2\theta}{dy^2} + \lambda^2 \theta = -\lambda^2 (\alpha_{\mathrm{r}} + \overline{\alpha}_{\mathrm{r}}). \tag{3.48}$$

The general solution to this linear ordinary differential equation is

$$\theta = A \sin(\lambda y) + B \cos(\lambda y) - (\alpha_{\mathrm{r}} + \overline{\alpha}_{\mathrm{r}}). \tag{3.49}$$

Applying the boundary conditions, one finds that

$$\begin{aligned} \theta(0) = 0: \qquad & B = \alpha_{\mathrm{r}} + \overline{\alpha}_{\mathrm{r}}, \\ \theta'(\ell) = 0: \qquad & A = B \tan(\lambda \ell), \end{aligned} \tag{3.50}$$

where $(\ )' = d(\ )/dy$. Thus, the elastic twist distribution becomes

$$\theta = (\alpha_{\mathrm{r}} + \overline{\alpha}_{\mathrm{r}}) \left[\tan(\lambda \ell) \sin(\lambda y) + \cos(\lambda y) - 1 \right]. \tag{3.51}$$

Since θ is now known, the spanwise lift distribution can be found using the relation

$$L' = qca(\alpha_{\mathrm{r}} + \theta). \tag{3.52}$$

It is important to note from the above expression for elastic twist that θ becomes infinite as $\lambda \ell$ approaches $\pi/2$. This phenomenon is called "torsional divergence" and depends on the numerical value of

$$\lambda = \sqrt{\frac{qcae}{GJ}}. \tag{3.53}$$

Thus, it is apparent that there exists a value of the dynamic pressure $q = q_D$, at which $\lambda \ell$ equals $\pi/2$, where the elastic twist theoretically becomes infinite. In practice, this static aeroelastic instability causes catastrophic failure of the wing structure. The value q_D is called the "divergence dynamic pressure" and is given by

$$q_D = \frac{GJ}{eca} \left(\frac{\pi}{2\ell} \right)^2. \tag{3.54}$$

Noting now that one can write

$$\lambda \ell = \frac{\pi}{2}\sqrt{\bar{q}} \tag{3.55}$$

with

$$\bar{q} = \frac{q}{q_D}, \tag{3.56}$$

the twist angle of the wing at the tip can be written as

$$\begin{aligned}
\theta(\ell) &= (\alpha_r + \bar{\alpha}_r)\left[\sec(\lambda \ell) - 1\right] \\
&= (\alpha_r + \bar{\alpha}_r)\left[\sec\left(\frac{\pi}{2}\sqrt{\bar{q}}\right) - 1\right],
\end{aligned} \tag{3.57}$$

where Eq. (3.47) can now be written as

$$\bar{\alpha}_r = \frac{cc_{mac}}{ae} - \frac{4\ell^2 Nmgd}{GJ\pi^2\bar{q}}. \tag{3.58}$$

Letting d be zero, so that $\bar{\alpha}_r$ becomes independent of $\bar{q}$, one can examine the behavior of $\theta(\ell)$ versus $\bar{q}$. Such a function is plotted in Fig. 3.13, where one sees that the tip twist angle goes to infinity as $\bar{q}$ approaches unity. Note that the character of the plot in Fig. 3.13 is quite similar to the prebuckling behavior of columns that have imperfections. It is of practical interest to note that the tip twist angle becomes sufficiently large to warrant concern about the structural integrity for dynamic pressures well below q_D.

Because this instability occurs at a dynamic pressure that is independent of the right-hand side of Eq. (3.48), as long as the right-hand side is nonzero, it seems possible that the divergence condition could be obtained from the homogeneous equilibrium equation

$$\frac{d^2\theta}{dy^2} + \lambda^2\theta = 0. \tag{3.59}$$

The general solution to this eigenvalue problem of the Sturm–Liouville type is

$$\theta = A\sin(\lambda y) + B\cos(\lambda y). \tag{3.60}$$

Applying the boundary conditions, one obtains

$$\begin{aligned}
\theta(0) &= 0: \quad B = 0, \\
\theta'(\ell) &= 0: \quad A\lambda \cos(\lambda \ell) = 0.
\end{aligned} \tag{3.61}$$

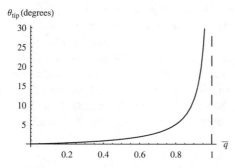

Figure 3.13   Plot of tip twist angle for wing versus $\bar{q}$ for $\alpha_r + \bar{\alpha}_r = 1°$.

If $A = 0$ in the last condition, there is no deflection; this is a so-called trivial solution. However, if $\lambda = 0$ the given general solution is not valid. Thus, the desired result is obtained when $\cos(\lambda\ell) = 0$. This is the "characteristic equation" with solutions given by

$$\lambda_n\ell = (2n - 1)\frac{\pi}{2} \qquad (n = 1, 2, \ldots). \tag{3.62}$$

These values are called "eigenvalues." Note that this set of λs corresponds to a set of dynamic pressures

$$q_n = (2n - 1)^2 \left(\frac{\pi}{2\ell}\right)^2 \frac{GJ}{eca} \qquad (n = 1, 2, \ldots). \tag{3.63}$$

The lowest of these values, q_1, is equal to the divergence dynamic pressure, q_D, previously obtained from the nonhomogeneous equilibrium equation. This result implies that there exist nontrivial solutions of the homogeneous equation for the elastic twist. In other words, even for cases in which the right-hand side of Eq. (3.48) is zero (when $\alpha_r + \overline{\alpha}_r = 0$), there is a nontrivial solution

$$\theta_n = A_n \sin(\lambda_n y) \tag{3.64}$$

for each of these discrete values of dynamic pressure. Since A_n is undetermined, the amplitude of θ_n is arbitrary, which means that the effective torsional stiffness is zero whenever the dynamic pressure $q = q_n$. The mode shape θ_1 is the divergence mode shape, which must not be confused with the twist distribution obtained from the nonhomogeneous equation.

It may be noted that if the elastic axis is upstream of the aerodynamic center then $e < 0$ and λ is imaginary in the preceding analysis. The characteristic equation for the divergence condition becomes $\cosh(|\lambda|\ell) = 0$. Because there is no real value of λ that satisfies this equation, the divergence phenomenon will not occur in this case.

3.2.3   Airload Distribution

It has been observed that the spanwise lift distribution can be determined as

$$L' = qca(\alpha_r + \theta), \tag{3.65}$$

where we recall from Eq. (3.51) that

$$\theta = (\alpha_r + \overline{\alpha}_r)\left[\tan(\lambda\ell)\sin(\lambda y) + \cos(\lambda y) - 1\right] \tag{3.66}$$

and where $\overline{\alpha}_r$ is given in Eq. (3.47). If the lifting surface is a wind tunnel model of a wing and is fastened to the wind tunnel wall, then the load factor, N, is equal to unity and α_r can be specified. The resulting computation of L' is straightforward.

If, however, the lifting surface represents half the wing surface of a flying vehicle, the computation of L' is not as direct. It may be noted that the constant $\overline{\alpha}_r$ is a function of N. Thus, for given value of α_r there will be a corresponding distribution of elastic twist and a particular airload distribution. This airload can be integrated over the vehicle to obtain the total lift, L. Recall that $N = L/W$, where W is the vehicle weight. It is thus apparent that the load factor, N, is related to the rigid angle of attack, α_r, through the elastic twist angle, θ. For this reason either of the two variables α_r and N can be specified; the other can then be obtained from the total lift L. Assuming a two-winged vehicle with all the lift being generated from the wings, one finds

$$L = 2\int_0^\ell L'\,dy. \tag{3.67}$$

Substituting for L' and α_r as given above yields

$$L = 2qca \int_0^\ell \left\{ \alpha_r + (\alpha_r + \overline{\alpha}_r) \left[\tan(\lambda\ell) \sin(\lambda y) + \cos(\lambda y) - 1 \right] \right\} dy$$

$$= 2qca\ell \left\{ (\alpha_r + \overline{\alpha}_r) \left[\frac{\tan(\lambda\ell)}{\lambda\ell} \right] - \overline{\alpha}_r \right\}. \tag{3.68}$$

Since $N = L/W$, the above expression can be divided by the vehicle weight to yield a relation for N in terms of α_r and $\overline{\alpha}_r$. This relation can then be solved simultaneously with the preceding expression for $\overline{\alpha}_r$, Eq. (3.47), in terms of α_r and N. In this manner $\overline{\alpha}_r$ can be eliminated, providing either a relation that expresses N in terms of α_r, given by

$$N = \frac{2GJ(\lambda\ell)^2 \left\{ ae\alpha_r + cc_{mac} \left[1 - \frac{\lambda\ell}{\tan(\lambda\ell)} \right] \right\}}{ae\ell \left\{ \frac{We\lambda\ell}{\tan(\lambda\ell)} + 2mg\ell d \left[1 - \frac{\lambda\ell}{\tan(\lambda\ell)} \right] \right\}} \tag{3.69}$$

or a relation that expresses α_r in terms of N,

$$\alpha_r = \frac{NW\ell e}{2GJ\lambda\ell \tan(\lambda\ell)} + \left[1 - \frac{\lambda\ell}{\tan(\lambda\ell)} \right] \left[\frac{Nmg\ell^2 d}{GJ(\lambda\ell)^2} - \frac{cc_{mac}}{ae} \right]. \tag{3.70}$$

These relations permit one to specify a constant α_r and find $N(q)$ or, alternatively, to specify a constant N and find $\alpha_r(q)$. One finds that $N(q)$ starts out at zero for $q = 0$. On the other hand, $\alpha_r(q)$ starts out at infinity for $q = 0$. The limiting values as $q \to q_D$ depend on the other parameters. These equations can be used to find the torsional deformation and the resulting airload distribution for a specified flight condition.

The calculation of the spanwise aeroelastic airload distribution is immensely practical and is used in industry in two separate ways. One way is to satisfy a requirement of the aerodynamicist or performance engineer who needs to know the total forces and moments on the flight vehicle as a function of its altitude and flight condition. In this instance the dynamic pressure q (and altitude or Mach number) and α_r are specified, and the load factor N or total lift L is computed using Eq. (3.69). A second requirement is that of the structural engineer, who must ensure the structural integrity of the lifting surface for a specified load factor N and flight condition. Such a specification is normally described by what is called a V–N diagram. For the conditions of given load factor and flight condition it is necessary for the structural engineer to know the airload distribution to conduct a subsequent loads and stress analysis. When q (and altitude or Mach number) and N are specified, α_r is then determined from Eq. (3.70). Knowing q, α_r, and N, one then uses Eq. (3.47) to find $\overline{\alpha}_r$. The torsional deformation, θ, then follows from Eq. (3.66), and the spanwise airload distribution follows from Eq. (3.65). From this, the torsional and bending moment distributions along the wing can be found, leading directly to the maximum stress in the wing, generally somewhere in the root cross section.

It may be observed that the overall effect of torsional flexibility on the unswept lifting surface is to significantly change the spanwise airload distribution. This effect can be seen as the presence of the elastic part of the lift coefficient, which is of course proportional to $\theta(y)$. Since this elastic torsional rotation will generally increase as the distance from the root (i.e., out along the span), so also will the resultant airload distribution. The net effect will depend on whether α_r or N is specified. If α_r is specified as in the case of a wall-mounted elastic wind tunnel model ($N = 1$) or as in performance computations, then the total lift will increase with the additional load appearing in the outboard region as shown in Fig. 3.14.

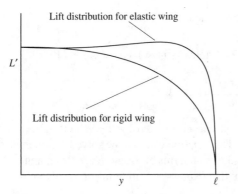

Figure 3.14 Rigid and elastic wing lift distributions holding α_r constant.

In the other case, when N is specified by the structural engineer, the total lift (area under L' versus y) is unchanged, as shown in Fig. 3.15. The addition of lift in the outboard region must be balanced by a decrease inboard. This is accomplished by decreasing α_r as the surface is made more flexible.

All the preceding equations for torsional divergence and airload distribution have been based on a strip-theory aerodynamic representation. It may be noted that a slight numerical improvement in their predictive capability can be obtained if the two-dimensional lift-curve slope, a, is replaced everywhere by the total (three-dimensional) lift-curve slope. Although there is little theoretical justification for this modification, it does alter the numerical results in the direction of the exact answer. Also, it is important to note that the lift distributions depicted in Figs. 3.14 and 3.15 cannot be generated with strip-theory aerodynamics, because strip theory fails to pick up the dropoff of the airload to zero at the wing tip caused by three-dimensional effects. A theory at least as sophisticated as lifting-line theory would have to be used to capture that effect.

3.2.4 Sweep Effects

To observe the effect of sweeping a wing aft or forward on the aeroelastic characteristics, it will be presumed that the swept geometry is obtained by rotating the surface about the root of the elastic axis as illustrated in Fig. 3.16. The aerodynamic reactions will

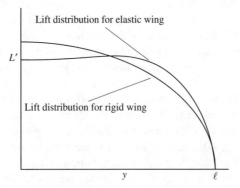

Figure 3.15 Rigid and elastic wing lift distributions holding total lift constant.

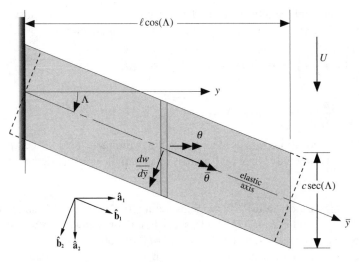

Figure 3.16   Schematic of swept wing (positive Λ).

depend on the angle of attack as measured in the streamwise direction as

$$\alpha = \alpha_r + \theta, \tag{3.71}$$

where θ is the change in the streamwise angle of attack caused by elastic deformation. To develop a kinematical relation for θ, we introduce the unit vectors $\hat{\mathbf{a}}_1$ and $\hat{\mathbf{a}}_2$, aligned with the y axis and the freestream, respectively. Another set of unit vectors, $\hat{\mathbf{b}}_1$ and $\hat{\mathbf{b}}_2$, are obtained by rotating $\hat{\mathbf{a}}_1$ and $\hat{\mathbf{a}}_2$ by the sweep angle Λ as shown in Fig. 3.16, so that $\hat{\mathbf{b}}_1$ is aligned with the elastic axis (i.e., the $\overline{y}$ axis). From Fig. 3.16 one sees that

$$\begin{aligned} \hat{\mathbf{b}}_1 &= \cos(\Lambda)\hat{\mathbf{a}}_1 + \sin(\Lambda)\hat{\mathbf{a}}_2, \\ \hat{\mathbf{b}}_2 &= -\sin(\Lambda)\hat{\mathbf{a}}_1 + \cos(\Lambda)\hat{\mathbf{a}}_2. \end{aligned} \tag{3.72}$$

It should be observed that the total rotation of the local wing cross-sectional frame caused by elastic deformation can be written as the combination of rotations caused by wing torsion, $\overline{\theta}$ about $\hat{\mathbf{b}}_1$, and wing bending, $dw/d\overline{y}$ about $\hat{\mathbf{b}}_2$, where w is the wing bending deflection (positive up). Now, θ is the component of this total rotation about $\hat{\mathbf{a}}_1$, that is,

$$\begin{aligned} \theta &= \left(\overline{\theta}\hat{\mathbf{b}}_1 + \frac{dw}{d\overline{y}}\hat{\mathbf{b}}_2\right) \cdot \hat{\mathbf{a}}_1 \\ &= \overline{\theta}\cos(\Lambda) - \frac{dw}{d\overline{y}}\sin(\Lambda). \end{aligned} \tag{3.73}$$

From this relation, it can be noted that, as the result of sweep, the effective angle of attack is altered by bending. This coupling between bending and torsion will affect both the static aeroelastic response of the wing in flight as well as the conditions under which divergence occurs. Also, it can be observed that, for combined bending and torsion of a swept elastic wing, the section in the direction of the streamwise airflow exhibits a change in camber, a higher-order effect that is here neglected.

To facilitate direct comparison with the previous unswept results, to the extent possible the same structural and aerodynamic notation will be retained as was used for the unswept planform. To determine the total elastic deflection two equilibrium equations are required, one for torsional moment equilibrium as in the unswept case and one for transverse force equilibrium (associated with bending). These equations can be written as

$$\frac{d}{d\overline{y}}\left(GJ\frac{d\overline{\theta}}{d\overline{y}}\right) = -qec\overline{a}\alpha - qc^2 c_{\mathrm{mac}} + Nmgd,$$

$$\frac{d^2}{d\overline{y}^2}\left(EI\frac{d^2 w}{d\overline{y}^2}\right) = qc\overline{a}\alpha - Nmg. \tag{3.74}$$

In these equilibrium equations $\overline{a}$ is used to denote the two-dimensional lift-curve slope of the swept surface. This aerodynamic derivative is related to the unswept derivative by

$$\overline{a} = a\cos(\Lambda) \tag{3.75}$$

for moderate- to high-aspect-ratio surfaces. Substituting for $\overline{a}$, $\alpha = \alpha_r + \theta$, and in turn the dependence of θ on $\overline{\theta}$ and w from Eq. (3.73), specializing for spanwise uniformity so that GJ and EI are constants, and letting $(\ )'$ denote $d(\ )/d\overline{y}$, one obtains two coupled, ordinary differential equations for torsion and bending given by

$$\overline{\theta}'' + \frac{qeca}{GJ}\overline{\theta}\cos^2(\Lambda) - \frac{qeca}{GJ}w'\sin(\Lambda)\cos(\Lambda) = -\frac{1}{GJ}[qeca\alpha_r\cos(\Lambda)$$

$$+ qc^2 c_{\mathrm{mac}} - Nmgd],$$

$$w'''' + \frac{qca}{EI}w'\sin(\Lambda)\cos(\Lambda) - \frac{qca}{EI}\overline{\theta}\cos^2(\Lambda) = \frac{1}{EI}[qca\alpha_r\cos(\Lambda) - Nmg]. \tag{3.76}$$

Since the surface is built-in at the root and free at the tip the following boundary conditions must be imposed on the solution:

$$\begin{array}{llll}
\overline{y} = 0: & \overline{\theta} = 0 & \text{(zero torsional rotation)}, \\
& w = 0 & \text{(zero deflection)}, \\
& w' = 0 & \text{(zero bending slope)}, \\
\overline{y} = \ell: & \overline{\theta}' = 0 & \text{(zero twisting moment)}, & (3.77) \\
& w'' = 0 & \text{(zero bending moment)}, \\
& w''' = 0 & \text{(zero shear force)}.
\end{array}$$

Bending–torsion coupling is exhibited in Eqs. (3.76) through the term involving w in the torsion equation and through the term involving $\overline{\theta}$ in the bending equation.

There are two special cases of interest in which the coupling either vanishes or is very much simplified, so that one can solve the equations analytically. The first is for the case of vanishing sweep in which the uncoupled torsion equation (the first of Eqs. 3.76) is the same as previously discussed and clearly leads to solutions for either the torsional divergence condition or the torsional deformation and airload distribution, as discussed in Sections 3.2.2 and 3.2.3, respectively. In the latter case, once the torsional deformation is obtained, the solution for $\theta = \overline{\theta}$ can be substituted into the bending equation (the second of Eqs. 3.76). Simple integration of the resulting ordinary differential equation leads to the shear force, bending moment, bending slope, and bending deflection.

A second special case occurs when $e = 0$. In this case, torsional divergence does not take place, and a polynomial solution for $\overline{\theta}$ can be found from the $\overline{\theta}$ equation and substituted into the bending equation. This leads to a fourth-order ordinary differential equation for w with a polynomial forcing function; note that the $\overline{\theta}$ terms are now part of that forcing function. This equation can be solved for the bending deflection, but the solution is not straightforward. Alternatively, to solve this equation for a divergence condition, one needs only the homogeneous part, which can be written as a third-order equation in $\zeta = w'$, namely,

$$\zeta''' + \frac{qca}{EI}\zeta \sin(\Lambda)\cos(\Lambda) = 0. \tag{3.78}$$

For the clamped–free boundary conditions $\zeta(0) = \zeta'(\ell) = \zeta''(\ell) = 0$, this equation has a known analytical solution that yields a divergence dynamic pressure of

$$q_D = -6.32970\frac{EI}{ac\ell^3 \sin(\Lambda)\cos(\Lambda)}. \tag{3.79}$$

[handwritten annotation: EA @ AC / No tailoring / e = o, K = o]

The minus sign implies that this bending divergence instability only takes place for forward-swept wings, that is, where $\Lambda < 0$.

Examination of Eqs. (3.76) illustrates that there are two ways in which the sweep influences the aeroelastic behavior. One is the loss of aerodynamic effectiveness as exhibited by the change in the second term of the torsion equations from

$$\frac{qeca}{GJ}\bar{\theta} \quad \text{to} \quad \frac{qeca}{GJ}\bar{\theta}\cos^2(\Lambda). \tag{3.80}$$

Note that this effect is independent of the direction of sweep. The second effect is the influence of bending slope on the effective angle of attack (see Eq. 3.73), which leads to bending–torsion coupling. This coupling has a strong influence on both divergence and load distribution. The total effect of sweep depends strongly on whether the surface is swept back or forward. This can be illustrated by its influence on the divergence dynamic pressure, q_D, as shown in Fig. 3.17. It is apparent that forward sweep causes the surface to be more susceptible to divergence whereas backward sweep increases the divergence dynamic pressure. Indeed, a small amount of backward sweep (for the idealized case under consideration, depending on e/ℓ and GJ/EI, only $5°$ or $10°$) can cause the divergence dynamic pressure to become infinite, thus eliminating the instability altogether. Some specific cases are discussed later in this section in conjunction with an approximate solution of the governing equations.

The overall effect of sweep on the aeroelastic load distribution also strongly depends on whether the surface is swept forward or aft. This is illustrated in Fig. 3.18, which shows spanwise load distributions for an elastic surface for which the total lift (or N) is held constant by adjusting α_r. From a structural loads standpoint it is apparent that the root bending moment is significantly greater for forward sweep than for backward sweep at a given value of total lift.

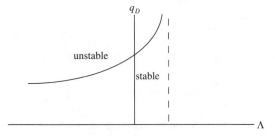

Figure 3.17 Divergence dynamic pressure versus Λ.

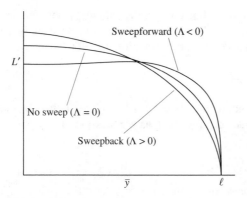

Figure 3.18   Lift distribution for positive, zero, and negative Λ.

The primary motivation for sweeping a lifting surface is to improve the vehicle performance through drag reduction, although some loss in lifting capability may be experienced.
However, the above aeroelastic considerations can have a significant impact on design de-
cisions. From an aeroelastic standpoint, forward sweep exacerbates divergence instability
and increases structural loads whereas backward sweep can alleviate these concerns. The
advent of composite lifting surfaces has enabled the use of bending–twist elastic coupling to
passively stabilize forward sweep, making it possible to use forward-swept wings. Indeed,
the X-29 could not have been flown without some means to stabilize the wings against
divergence. We will discuss this further below under aeroelastic tailoring.

Exact Solution for Bending–Torsion Divergence

Extraction of the analytical solution of the above set of coupled ordinary differential equations, Eqs. (3.76), is very complicated. The exact analytical solution is most
easily obtained by first converting the coupled set of equations into a single equation governing the elastic component of the angle of attack. For calculation of only the divergence
dynamic pressure, one can consider just the homogeneous parts of Eqs. (3.76):

$$\bar{\theta}'' + \frac{qeca}{GJ}\bar{\theta}\cos^2(\Lambda) - \frac{qeca}{GJ}w'\sin(\Lambda)\cos(\Lambda) = 0,$$

$$w'''' + \frac{qca}{EI}w'\sin(\Lambda)\cos(\Lambda) - \frac{qca}{EI}\bar{\theta}\cos^2(\Lambda) = 0. \tag{3.81}$$

To get a single equation one differentiates the first equation with respect to $\bar{y}$ and multiplies it
by $\cos(\Lambda)$. From this modified first equation, one subtracts $\sin(\Lambda)$ times the second equation,
replacing $\bar{\theta}\cos(\Lambda) - w'\sin(\Lambda)$ with θ, to obtain

$$\theta''' + \frac{qeca}{GJ}\cos^2(\Lambda)\theta' + \frac{qca}{EI}\sin(\Lambda)\cos(\Lambda)\theta = 0. \tag{3.82}$$

Introducing a dimensionless axial coordinate $\eta = \bar{y}/\ell$, the above equation can be written as

$$\theta''' + \frac{qeca\ell^2}{GJ}\cos^2(\Lambda)\theta' + \frac{qca\ell^3}{EI}\sin(\Lambda)\cos(\Lambda)\theta = 0, \tag{3.83}$$

where ( )$'$ now denotes $d(\ )/d\eta$. The boundary conditions can be derived from Eqs. (3.77) as

$$\theta(0) = \theta'(1) = \theta''(1) + \frac{qeca\ell^2}{GJ}\cos^2(\Lambda)\theta(1) = 0. \tag{3.84}$$

Here the first of Eqs. (3.81) and the final boundary condition from Eqs. (3.77) are used to derive the third boundary condition.

The exact solution for Eqs. (3.83) and (3.84) has been obtained by Diederich and Budiansky (1948). Its behavior is quite complex, having multiple branches, and it is not easily used in a design context. However, a simple approximation of one branch is presented next and compared with plots of the exact solution.

Approximate Solution for Bending–Torsion Divergence

In view of the complexity of the exact solution, it is fortunate that there are various approximate methods for treating such equations, one of which is the application of the method of Ritz to the virtual work principle (see Section 2.4). In this special case, the kinetic energy is zero, and the resulting algebraic equations are a special case of the generalized equations of motion (see Section 2.1.6), termed generalized equations of equilibrium. Determination of such an approximate solution is left as an exercise for the reader; see Problems 11–15.

Here we consider instead an approximation of one branch of the analytical solution for the bending–torsion divergence problem. Fortunately, the most important branch from a physical point of view behaves quite simply. Indeed, if one defines

$$\tau = \frac{qeca\ell^2}{GJ}\cos^2(\Lambda),$$

$$\beta = \frac{qca\ell^3}{EI}\sin(\Lambda)\cos(\Lambda),$$

(3.85)

then, as shown by Diederich and Budiansky (1948), the divergence boundary can be approximately represented within a certain range in terms of a straight line

$$\tau_D = \frac{\pi^2}{4} + \frac{3\pi^2}{76}\beta_D.$$

(3.86)

Note that for a wing rigid in bending we have $\beta_D = 0$, and thus $\tau_D = \frac{\pi^2}{4}$, which is the exact solution for pure torsional divergence. Also, for a torsionally rigid wing we have $\tau_D = 0$, and thus $\beta_D = -19/3$, which is very close to -6.3297, the exact solution for bending divergence. For the cases in between the error is quite small. It is very important to note that the sign of τ is driven by the sign of e, whereas the sign of β is driven by the sign of Λ. The approximate solution in Eq. (3.86) is plotted along with some of the exact solution branches in Fig. 3.19. Note the excellent agreement between the straight line approximation and the exact solution near the origin. Also note that the intersections of the solution with the τ_D axis (where $\beta_D = 0$) coincide with the squares of the roots previously obtained in Section 3.2.2, Eq. (3.62), as $(2n-1)^2\pi^2/4$ for $n = 1, 2, \ldots, \infty$ (i.e., $\pi^2/4, 9\pi^2/4, \ldots$).

A somewhat more convenient way of depicting the behavior of the divergence dynamic pressure is to plot τ_D versus a parameter that depends only on the configuration. This can be accomplished by introducing the dimensionless parameter r, given by

$$r \equiv \frac{\beta}{\tau} = \frac{\ell}{e}\frac{GJ}{EI}\tan(\Lambda),$$

(3.87)

which can be positive, negative, or zero. Equation (3.86) can then be written as

$$\tau_D = \frac{\pi^2}{4} + \frac{3\pi^2 r}{76}\tau_D.$$

(3.88)

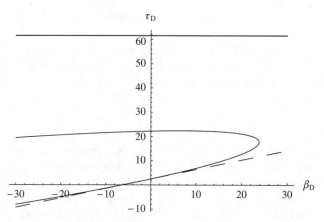

Figure 3.19 τ_D versus β_D for coupled bending–torsion divergence; solid lines (exact solution) and dashed line (Eq. 3.86).

Thus, one can solve for τ_D, such that

$$\tau_D = \frac{\pi^2}{4\left(1 - \frac{3\pi^2 r}{76}\right)},\qquad (3.89)$$

or alternatively for q_D, equal to

No tailoring (K=0)

$$q_D = \frac{GJ\pi^2}{4eca\ell^2\cos^2(\Lambda)\left[1 - \frac{3\pi^2}{76}\frac{\ell}{e}\frac{GJ}{EI}\tan(\Lambda)\right]}.\qquad (3.90)$$

Several branches of the exact solution of Eqs. (3.83) and (3.84) for the smallest absolute values of τ_D versus r are plotted as solid lines in Fig. 3.20. It is noted that there is at least one branch in all quadrants but the third, and there is only one branch in the fourth quadrant. The approximate solutions for τ_D versus r from Eq. (3.89) are plotted as dashed hyperbolae in the first, second, and fourth quadrants. Moreover, as r becomes large the solution in the fourth quadrant asymptotically approaches the parabola $\tau_D = -27r^2/4$, also shown as a dashed curve. We note that, as with Fig. 3.19, the intersections of the roots with the τ_D axis are $\pi^2/4$, $9\pi^2/4$, $25\pi^2/4$, etc. The configuration of any wing fixes the value of r. For positive e one considers only positive values of τ_D. Thus, one starts from zero and proceeds

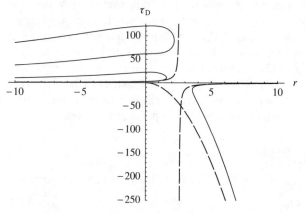

Figure 3.20 τ_D versus r for coupled bending–torsion divergence; solid lines (exact solution) and dashed lines (Eq. 3.89 and $\tau_D = -27r^2/4$ in fourth quadrant).

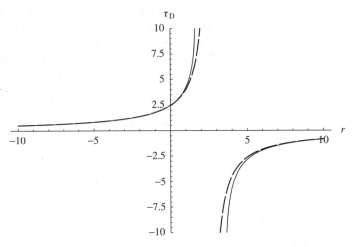

Figure 3.21 τ_D versus r for coupled bending–torsion divergence; solid lines (exact solution) and dashed lines (Eq. 3.89).

in the positive τ_D direction on this plot (i.e., at constant r) to find the first intersection with a solid line. This is the dynamic pressure parameter τ_D at which divergence occurs. In Fig. 3.21 a blowup of these results in a more practical range is shown. It is easily seen that the dashed lines in the first and second quadrants are very close to the solid ones when $r < 1.5$. We note that when $e < 0$ a negative value of τ_D leads to a positive value of q_D. In this case, one should proceed along a line of constant r in the negative τ_D direction.

It is interesting to note that the approximate solution, despite its close proximity to the exact solution, exhibits a qualitatively different behavior mathematically. The approximate solution exhibits an asymptotic behavior, with τ_D tending to plus infinity from the left and to minus infinity from the right at the value of r that causes the denominator to vanish, namely, when $r = 76/(3\pi^2) = 2.56680$. If the approximate solution were exact, mathematically it would mean that divergence is not possible at that value of r. Moreover, physically it would mean that divergence is not possible for $e > 0$ and $r \geq 76/(3\pi^2)$ [or for $e < 0$ and $r \leq 76/(3\pi^2)$]. Actually, however, the exact solution is of the "limit point" variety. For $e > 0$ this means that divergence occurs for small and positive values of r. Moreover, as r is increased in the first quadrant, τ_D also increases until a certain point is reached at which two things happen: (1) Above this value of τ_D, the curve turns back to the left instead of reaching an asymptote; and (2) any slight increase in r beyond this point will cause the solution to jump to a higher branch. This point is called a limit point. On the main branch of the curve in the first quadrant, for example, the limit point is at $r = 1.59768$ and $\tau_D = 10.7090$. It can be seen from the plot in Fig. 3.20 that any slight increase in r will cause the solution to jump from the lower branch, where its value is 10.7090, to a higher branch, where its value is 66.8133, at which point τ_D is rapidly increasing with r. So, although there is no value of r that will result in an infinite exact value of the divergence dynamic pressure, practically speaking, divergence in the vicinity of the limit point value of r is all but eliminated. Thus, it is sufficient for practical purposes to say that divergence is not possible near those points where the approximate solution blows up, and one may regard the approximate solution as being sufficiently close to the exact solution for design purposes. The limit point in the fourth quadrant is appropriate for the situation in which $e < 0$, namely, when the aerodynamic center is behind the elastic axis. There the exact limit point is at $r = 3.56595$ and $\tau_D = -14.8345$. Note that the negative values of e and τ_D yield a positive q_D. It is left to the reader as an exercise to explore this possibility further; see Problem 17.

Although there are qualitative differences as noted between the exact and approximate
solutions, within the practical range of interest, the above linear approximation of the
divergence boundary in terms of τ_D and β_D is numerically quite accurate and leads to
a very simple expression for the divergence dynamic pressure in terms of the structural
stiffnesses, e/ℓ, and the sweep angle, that is, Eq. (3.90). This approximate formula can be
used in design to explore the behavior of the divergence dynamic pressure as a function of
the various configuration parameters therein. For the purpose of displaying results for the
divergence dynamic pressure when $e > 0$, it is convenient to normalize q_D with its value at
zero sweep angle, namely,

$$q_{D_0} = \frac{\pi^2 GJ}{4eca\ell^2}, \tag{3.91}$$

so that

$$\frac{q_D}{q_{D_0}} = \frac{1 + \tan^2(\Lambda)}{1 - \frac{3\pi^2}{76}\frac{\ell}{e}\frac{GJ}{EI}\tan(\Lambda)}. \tag{3.92}$$

Thus, for a wing structural design with given values of $e > 0$, GJ, EI, and ℓ, there are
multiple values of sweep angle Λ for which the divergence dynamic pressure goes to infinity,
implying that divergence is not possible at those values of Λ. There are evidently some values
of Λ that make the numerator term $\tan(\Lambda)$ infinite, whereas there are other values that make
the denominator vanish. Therefore, within the principal range of Λ, we can surmise that
divergence will only take place for cases in which $|\Lambda| \neq 90°$ and that $3\pi^2 r \neq 76$; that is,
divergence is not possible for $e > 0$ unless $-90° < \Lambda < \Lambda_\infty$ where

$$\tan(\Lambda_\infty) = \frac{76 EIe}{3\pi^2 GJ\ell}. \tag{3.93}$$

Thus, Eq. (3.92) can be written as

$$\frac{q_D}{q_{D_0}} = \frac{1 + \tan^2(\Lambda)}{1 - \frac{\tan(\Lambda)}{\tan(\Lambda_\infty)}}. \tag{3.94}$$

In other words, one avoids divergence by choosing $\Lambda \geq \Lambda_\infty$, and the divergence dynamic
pressure drops drastically as Λ is decreased below Λ_∞. Because Λ_∞ is likely to be small, this
frequently means that backswept wings are free of divergence, and that divergence dynamic
pressure drops drastically for forward-swept wings. Because Λ_∞ is the asymptotic value
of Λ from the approximate solution, which is greater than the limit point value of Λ from
the exact solution, we may surmise that the approximate solution provides a conservative
design. Figure 3.22 shows the behavior of divergence dynamic pressure for an elastically
uncoupled wing with $GJ/EI = 1.0$ and $e/\ell = 0.02$. The plot, as expected, passes through
unity when the sweep angle is zero. Since Λ_∞ is very small for this case, the divergence
dynamic pressure goes to infinity for a very small positive value of sweep angle. Thus,
even a small backward sweep angle can make divergence impossible. Figure 3.23 shows
the result of decreasing GJ/EI to 0.2 and holding e/ℓ constant. Because Λ_∞ increases,
the wing must be swept back further than in the previous case to avoid divergence.

Aeroelastic Tailoring
Aeroelastic tailoring is the design of wings using the directional properties of composite materials to optimize aeroelastic performance. The concept of aeroelastic tailoring
is relatively new and came into the forefront during the design of forward-swept wings in
the 1980s. Equation (3.94) shows that q_D drops dramatically for forward-swept, untailored
wings. The low divergence speed was a major hurdle in the design of wings with forward

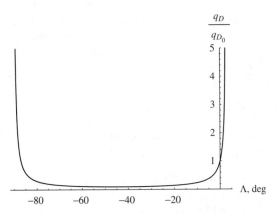

Figure 3.22   Normalized divergence dynamic pressure for an elastically uncoupled, swept wing with $GJ/EI = 1.0$ and $e/\ell = 0.02$.

sweep. As will be seen in this section, use of composite materials can help remove the disadvantages of forward sweep. Presently, aeroelastic tailoring is an integral part of composite wing designs and can be used to provide optimum performance.

Composite materials are anisotropic, which implies different material characteristics (such as stiffness) in different directions. A simple beam model is quite helpful in developing an understanding of the behavior of composite wings. Such models may exhibit bending–torsion elastic coupling. Analysis of beams with elastic coupling is a bit more involved, but it leads to results that are quite helpful.

Let us introduce such coupling in our beam equations. For anisotropic beams with bending–torsion coupling, the "constitutive law" (i.e., the relation between cross-sectional stress resultants and generalized strains) changes from

$$
\left\{ \begin{array}{c} T \\ M \end{array} \right\} = \left[\begin{array}{cc} GJ & 0 \\ 0 & EI \end{array} \right] \left\{ \begin{array}{c} \overline{\theta}' \\ w'' \end{array} \right\} \tag{3.95}
$$

to

$$
\left\{ \begin{array}{c} T \\ M \end{array} \right\} = \left[\begin{array}{cc} GJ & -K \\ -K & EI \end{array} \right] \left\{ \begin{array}{c} \overline{\theta}' \\ w'' \end{array} \right\}, \tag{3.96}
$$

where K is the bending–torsion coupling stiffness (having the same dimensions as EI and GJ) and $(\ )'$ indicates the derivative with respect to $\overline{y}$. A positive value of K means that a

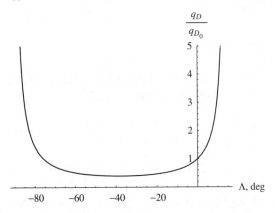

Figure 3.23   Normalized divergence dynamic pressure for an elastically uncoupled, swept wing with $GJ/EI = 0.2$ and $e/\ell = 0.02$.

positive bending deflection will be accompanied by a nose-up twist of the wing, which is normally destabilizing for cases with the elastic axis behind the aerodynamic center.

Using the coupled constitutive law, the equations of equilibrium become

$$(GJ\overline{\theta}' - Kw'')' = -qec\overline{a}\alpha - qc^2 c_{mac} + Nmgd,$$

$$(EIw'' - K\overline{\theta}')'' = qc\overline{a}\alpha - Nmg. \tag{3.97}$$

Let us again consider a wing that is clamped at the root and free at the tip, so that the boundary conditions that must be imposed on the solution are

$$\overline{y} = 0: \qquad \overline{\theta} = 0 \qquad \text{(zero torsional rotation)},$$
$$w = 0 \qquad \text{(zero deflection)},$$
$$w' = 0 \qquad \text{(zero bending slope)},$$
$$\overline{y} = \ell: \qquad T = 0 \qquad \text{(zero twisting moment)},$$
$$M = 0 \qquad \text{(zero bending moment)},$$
$$M' = 0 \qquad \text{(zero shear force)}. \tag{3.98}$$

For composite beams the offsets d and e may be defined in a manner similar to the way they were defined for isotropic beams: d is the distance from the $\overline{y}$-axis to the cross-sectional mass centroid, positive when the mass cetroid is toward the leading edge from the $\overline{y}$-axis; and e is the distance from the $\overline{y}$-axis to the aerodynamic center, positive when the aerodynamic center is toward the leading edge from the $\overline{y}$-axis. However, for composite beams the $\overline{y}$-axis must have different properties from those it possesses for isotropic beams, and the term "elastic axis" has a different meaning. Recall that for a spanwise uniform isotropic beam, the elastic axis is along the $\overline{y}$-axis and is the locus of cross-sectional shear centers; transverse forces acting through this axis do not twist the beam. For spanwise uniform composite beams with bending-twist coupling no axis can be defined as the locus of a cross-sectional property through which transverse shear forces can act without twisting the beam. For such beams we must place the $\overline{y}$-axis along the locus of generalized shear centers, a point in the cross-section at which transverse shear forces are structurally decoupled from the twisting moment; see Hodges (2006). Although transverse shear forces acting at the $\overline{y}$-axis do not *directly* induce twist, the bending moment induced by the shear force will still induce twist when $K \neq 0$.

We can now write the homogeneous part of the equations of equilibrium as

$$\overline{\theta}'' - \frac{K}{GJ}w''' + \frac{qeca}{GJ}\theta\cos(\Lambda) = 0$$

$$w'''' - \frac{K}{EI}\overline{\theta}''' - \frac{qca}{EI}\theta\cos(\Lambda) = 0 \tag{3.99}$$

Differentiating the first equation with respect to $\overline{y}$ and transforming the set of equations so that they are uncoupled in the highest derivative terms $\overline{\theta}'''$ and w'''', we obtain

$$\overline{\theta}''' + \frac{EI\,GJ}{EI\,GJ - K^2}\frac{qeca}{GJ}\theta'\cos(\Lambda) - \frac{K\,EI}{EI\,GJ - K^2}\frac{qca}{EI}\theta\cos(\Lambda) = 0$$

$$w'''' + \frac{K\,GJ}{EI\,GJ - K^2}\frac{qeca}{GJ}\theta'\cos(\Lambda) - \frac{EI\,GJ}{EI\,GJ - K^2}\frac{qca}{EI}\theta\cos(\Lambda) = 0 \tag{3.100}$$

Multiplying the first equation by $\cos(\Lambda)$ and the second equation by $\sin(\Lambda)$ and subtracting the second equation from the first, we obtain a single equation in terms of

Note that the aeroelastic divergence problem with structural coupling has the same mathematical form as the problem without coupling, an approximate solution of which is given in the previous section. One can see that the parameters τ and β can be redefined as

$$\tau = \frac{EI\,GJ}{EI\,GJ - K^2}\frac{qeca\ell^2}{GJ}\cos^2(\Lambda)\left[1 - \frac{K}{EI}\tan(\Lambda)\right],$$

$$\beta = \frac{EI\,GJ}{EI\,GJ - K^2}\frac{qca\ell^3}{EI}\sin(\Lambda)\cos(\Lambda)\left[1 - \frac{K}{GJ}\frac{1}{\tan(\Lambda)}\right],$$
(3.103)

and, again, the divergence boundary can be expressed approximately in terms of the line

$$\tau_D = \frac{\pi^2}{4} + \frac{3\pi^2}{76}\beta_D.$$
(3.104)

Using the expressions for the parameters in the equation of the divergence boundary, we have

$$q_D = \frac{\pi^2}{4}\frac{EI\,GJ - K^2}{EI\,GJ}\frac{GJ}{eca\ell^2\cos^2(\Lambda)}\frac{1}{\left\{1 - \frac{K}{EI}\tan(\Lambda) - \frac{3\pi^2}{76}\frac{\ell}{e}\frac{GJ}{EI}\left[\tan(\Lambda) - \frac{K}{GJ}\right]\right\}}.$$
(3.105)

One can simplify the above by introducing the dimensionless parameter

$$\kappa = \frac{K}{\sqrt{EI\,GJ}}$$
(3.106)

so that

$$q_D = \frac{\pi^2 GJ(1 - \kappa^2)}{4eca\ell^2\cos^2(\Lambda)\left\{1 - \kappa\sqrt{\frac{GJ}{EI}}\tan(\Lambda) - \frac{3\pi^2}{76}\frac{\ell}{e}\frac{GJ}{EI}\left[\tan(\Lambda) - \kappa\sqrt{\frac{EI}{GJ}}\right]\right\}}.$$
(3.107)

With Eq. (3.107) one may determine the divergence dynamic pressure with sufficient accuracy to ascertain its trends versus sweep angle Λ and elastic coupling parameter κ. The formula shows that there is a strong relationship between these two quantities.

To illustrate the utility of the above analysis, let us first normalize q_D with the value it would have at zero sweep angle and zero coupling, namely, q_{D_0}, so that

$$\frac{q_D}{q_{D_0}} = \frac{(1 - \kappa^2)[1 + \tan^2(\Lambda)]}{1 - \kappa\sqrt{\frac{GJ}{EI}}\tan(\Lambda) - \frac{3\pi^2}{76}\frac{\ell}{e}\frac{GJ}{EI}\left[\tan(\Lambda) - \kappa\sqrt{\frac{EI}{GJ}}\right]}.$$
(3.108)

The denominator's vanishing corresponds to infinite divergence dynamic pressure, and crossing this "boundary" means crossing from a regime in which divergence exists to one in which it does not. Setting the denominator to zero and solving for the tangent of the sweep angle, one obtains

$$\tan(\Lambda_\infty) = \frac{1 + \frac{3\pi^2}{76}\sqrt{\frac{GJ}{EI}}\frac{\ell}{e}\kappa}{\frac{3\pi^2}{76}\frac{GJ}{EI}\frac{\ell}{e} + \sqrt{\frac{GJ}{EI}}\kappa},$$
(3.109)

where Λ_∞ is the sweep angle at which the divergence dynamic pressure goes to infinity. With this definition, one can rewrite Eq. (3.108) as

$$\frac{q_D}{q_{D_0}} = \frac{(1 - \kappa^2)[1 + \tan^2(\Lambda)]}{\left(1 + \frac{3\pi^2}{76}\sqrt{\frac{GJ}{EI}}\frac{\ell}{e}\kappa\right)\left[1 - \frac{\tan(\Lambda)}{\tan(\Lambda_\infty)}\right]}$$
(3.110)

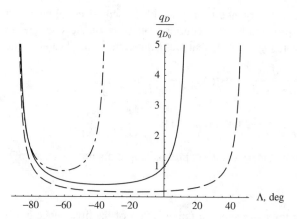

Figure 3.24 Normalized divergence dynamic pressure for an elastically coupled, swept wing with $GJ/EI = 0.2$ and $e/\ell = 0.02$; $\kappa = -0.4$ (dots and dashes), $\kappa = 0$ (solid lines), $\kappa = 0.4$ (dashed lines).

Again, for positive e, divergence is possible only if $-90° < \Lambda < \Lambda_\infty$. Thus, because of the presence of κ as an additional design parameter, the designer can at least partially compensate for the destabilizing effect of forward sweep by appropriately choosing $\kappa < 0$, which for an increment of upward bending of the wing provides an increment of nose-down twisting. There is a limit to how much coupling can be achieved, however, as typically, $|\kappa| < 0.6$.

There are two main differences in isotropic wing design versus design with composite wings. First, it is possible to achieve a much wider range of values for GJ/EI. Second, and significantly more powerful, is the fact that composite wings can be designed with nonzero values of κ. From Eq. (3.109), the value of Λ_∞ is decreased as κ is decreased, which means that the range of Λ over which divergence occurs is decreased. To confirm this and our earlier statement about positive κ being destabilizing, Fig. 3.24 shows results for $\kappa = -0.4$, 0, and 0.4. It is clear that one can sweep a composite wing forward and still avoid divergence with a proper choice (i.e., a sufficiently large and negative value) of κ. Since forward sweep has advantages for the design of highly maneuverable aircraft, this is a result of practical importance. The sweep angles at which divergence becomes impossible, Λ_∞, are also somewhat sensitive to GJ/EI and e/ℓ as shown in Figs. 3.25 and 3.26. Evidently, one may design divergence-free, forward-swept wings with larger sweep angles by decreasing torsional stiffness relative to bending stiffness and by decreasing e/ℓ.

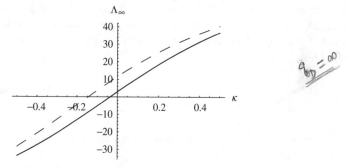

Figure 3.25 Sweep angle for which divergence dynamic pressure is infinite for a wing with $GJ/EI = 0.5$; solid line is for $e/\ell = 0.01$; dashed line is for $e/\ell = 0.04$.

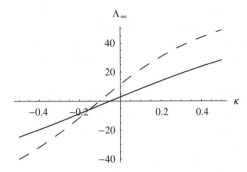

Figure 3.26 Sweep angle for which divergence dynamic pressure is infinite for a wing with $e/\ell = 0.02$; solid line is for $GJ/EI = 1.0$; dashed line is for $GJ/EI = 0.25$.

3.3   Epilogue

In this chapter we have considered divergence and aileron reversal of simple wind tunnel models, torsional divergence and load redistribution in flexible beam representations of lifting surfaces, the effects of sweep on coupled bending-torsion divergence, and the role of aeroelastic tailoring. In all these cases, the inertial loads are inconsequential and have thus been neglected. In Chapter 4, inertial loads are introduced into the aeroelastic analysis of flight vehicles, and the flutter problem is explored.

Problems

1. Consider a rigid, wind tunnel model of a uniform wing, which is pivoted in pitch about the mid-chord and elastically restrained in pitch by a linear spring with spring constant of 225 lb/in mounted at the trailing edge. The model has a symmetric airfoil, a span of 3 ft, and a chord of 6 in. The total lift-curve slope is 6 per rad. The aerodynamic center is located at the quarter-chord, and the mass centroid is at the mid-chord.
   (a) Calculate the divergence dynamic pressure at sea level.
   (b) Calculate the divergence airspeed at sea level.

   Answers: (a) $q_D = 150$ lb/ft^2; (b) $U_D = 355$ ft/s

2. For the model of Problem 1, for a dynamic pressure of 30 lb/ft^2, compute the percentage change in lift caused by the aeroelastic effect.

   Answer: 25%

3. For the model of Problem 1, propose design changes in the support system that would double the divergence dynamic pressure by
   (a) changing the stiffness of the restraining spring
   (b) relocating the pivot point

   Answers: (a) $k = 450$ lb/in; (b) $x_O = 2.513$ in

4. For the model of Problem 1 as altered by the design changes of Problem 3, calculate the percentage change in lift caused by the aeroelastic effect for a dynamic pressure of 30 lb/ft^2, a weight of 3 lb, $\alpha_r = 0.5°$, and for
   (a) the design change of Problem 3a
   (b) the design change of Problem 3b

   Answers: (a) 11.11%; (b) 17.91%

5. Consider a strut-mounted wing similar to the one discussed in Section 3.1.3, except that the two springs may have different stiffnesses. Denoting the leading-edge spring constant by k_1 and the trailing-edge spring constant by k_2 and assuming that the aerodynamic center is at the quarter chord, show that divergence can be eliminated if $k_1/k_2 \geq 3$.

6. Using Excel or a similar tool, plot a family of curves that depicts the relationship of the aileron elastic efficiency, η, versus normalized dynamic pressure, $\bar{q} = q/q_D$, for various values of $R = q_R/q_D$ and $0 \leq \bar{q} \leq 1$. You should make two plots on the following scales to reduce confusion:
   (a) Plot $R < 1$ using axes $-3 < \eta < 3$.
   (b) Plot $R > 1$ using axes $-3 < \eta < 3$.

   Hint: Do not compute values for the cases where $1 < R < 1.1$; Excel does not handle these well and you may get confused. For some cases you may want to plot symbols only and nicely sketch the lines that form the curves.

   Answer the following questions: Where does aileron reversal occur? If you had to design a wing, what R would you try to match (or approach) and why? What happens when $q_R = q_D$? How does the efficiency change as q approaches q_R? Why do you think this happens? What other pertinent features can you extract from these plots? Explain how you came to these conclusions.

7. Consider a torsionally elastic ($GJ = 8000$ lb in^2) wind tunnel model of a uniform wing, the ends of which are rigidly fastened to the wind tunnel walls. The model has a symmetric airfoil, a span of 3 ft, and a chord of 6 in. The sectional lift-curve slope is 6 per rad. The aerodynamic center is located at the quarter-chord, and both the mass centroid and the elastic axis are at the mid-chord.
   (a) Calculate the divergence dynamic pressure at sea level.
   (b) Calculate the divergence airspeed at sea level.

   Answers: (a) $q_D = 162.46$ lb/ft^2; (b) $U_D = 369.65$ ft/s

8. For the model of Problem 7, propose design changes in the model that would double the divergence dynamic pressure by
   (a) changing the torsional stiffness of the wing
   (b) relocating the elastic axis

   Answers: (a) $GJ = 16,000$ lb in^2; (b) $x_{ea} = 2.25$ in

9. For the model of Problem 7, for a dynamic pressure of 30 lb/ft^2, compute the percentage increase in the sectional lift at mid-span caused by the aeroelastic effect.

   Answer: 28.094%

10. For the model of Problem 7, for a dynamic pressure of 30 lb/ft^2, compute the percentage increase in the total lift caused by the aeroelastic effect.

   Answer: 18.59%

11. Consider a swept clamped–free wing, as described in Section 3.2.4. The governing partial differential equations are given in Eqs. (3.76) and the boundary conditions in Eqs. (3.77). An approximate solution is sought for a wing with a symmetric airfoil, using a truncated set of assumed modes and the generalized equations of equilibrium – a specialized version of the generalized equations of motion for which all time-dependent terms are zero. Note that what is being asked for here is

equivalent to the application of the method of Ritz to the principle of virtual work; see Section 2.4. With the wing weight ignored, only structural and aerodynamic terms are involved. The structural terms of the generalized equations of equilibrium are based on the potential energy (here the strain energy) given by

$$P = \frac{1}{2} \int_0^\ell (EI w''^2 + GJ\bar{\theta}'^2) \, d\bar{y}$$

and the bending and torsion deformation is represented in terms of a truncated series, such that

$$w = \sum_{i=1}^{N_w} \eta_i \Psi_i(\bar{y}),$$

$$\bar{\theta} = \sum_{i=1}^{N_\theta} \phi_i \Theta_i(\bar{y}),$$

where N_w and N_θ are the numbers of assumed modes used to represent bending and torsion, respectively; η_i and ϕ_i are the generalized coordinates associated with bending and torsion, respectively; and Ψ_i and Θ_i are the assumed mode shapes for bending and torsion, respectively. Determine the potential energy in terms of the generalized coordinates using as assumed modes the uncoupled, clamped–free, free-vibration modes of torsion and bending. For torsion

$$\Theta_i = \sqrt{2} \sin\left[\frac{\pi \left(i - \frac{1}{2}\right) \bar{y}}{\ell} \right];$$

for bending, according to Eq. (2.251), Ψ_i is given as

$$\Psi_i = \cosh(\alpha_i \bar{y}) - \cos(\alpha_i \bar{y}) - \beta_i [\sinh(\alpha_i \bar{y}) - \sin(\alpha_i \bar{y})]$$

with α_i and β_i as given in Table 2.1.

12. Rework Problem 11, but for assumed modes, instead of using the expressions given therein, use

$$\Theta_i = \left(\frac{\bar{y}}{\ell}\right)^i,$$

$$\Psi_i = \left(\frac{\bar{y}}{\ell}\right)^{i+1},$$

keeping in mind that these functions are not orthogonal.

13. Referring back to either Problem 11 or 12, starting with the virtual work of the aerodynamic forces as

$$\overline{\delta W} = \int_0^\ell (L' \delta w + M' \delta\bar{\theta}) \, d\bar{y},$$

where L' and M' are the sectional lift and pitching moment expressions used to develop Eqs. (3.76), and using the given deformation modes, find the generalized forces $\Xi_i, i = 1, 2, \ldots, N = N_w + N_\theta$. As discussed in the text, generalized forces are the coefficients of the variations of the generalized coordinates in the virtual work expression.

Hint: Neglecting the weight terms on the right-hand sides of Eqs. 3.74, one finds that L' is the right-hand side of the second of those equations, whereas M' is the negative of the right-hand side of the first and equal to eL'.

14. Referring back to Problems 13 and either 11 or 12, determine the generalized equations of equilibrium in the form

$$[K]\{\xi\} = \bar{q}\{[A]\{\xi\} + \{\Xi_0\}\},$$

where $\bar{q}$ is the dimensionless dynamic pressure given by q/q_{D_0}, q_{D_0} is the torsional divergence dynamic pressure of the unswept clamped–free wing, given by Eq. (3.54), $\{\xi\}$ is the column matrix of all unknowns $\bar{\eta}_i = \eta_i/\ell$, $i = 1, 2, \ldots, N_w$, and ϕ_i, $i = 1, 2, \ldots, N_\theta$, and $\{\Xi_0\}$ is a column matrix containing the parts of the aerodynamic generalized forces that do not depend on any unknowns. The $N \times N$ matrices $[K]$ and $[A]$ are the stiffness and aerodynamic matrices, respectively. If Problem 11 is the basis for solution, then the stiffness matrix $[K]$ is diagonal since the normal modes used to represent the wing structural behavior are orthogonal with respect to the stiffness properties of the wing.

15. Referring back to Problem 14, perform the following numerical studies:
    (a) *Divergence*: To determine the divergence dynamic pressure, write the *homogeneous* generalized equations of equilibrium in the form

$$\frac{1}{q}\{\xi\} = [K]^{-1}[A]\{\xi\},$$

which is obviously an eigenvalue problem with $1/\bar{q}$ as the eigenvalue. After you solve the eigenvalue problem, the largest $1/\bar{q}$ provides the lowest dimensionless critical divergence dynamic pressure $\bar{q}_D = q_D/q_{D_0}$ at the sweep angle under consideration. By numerical experimentation, determine how many modes are needed for each of w and $\bar{\theta}$ to obtain the divergence dynamic pressure to within plotting accuracy. Plot the divergence dynamic pressure versus sweep angle for a range of values for the sweep angle $-45° \leq \Lambda \leq 45°$ and values of the dimensionless parameters e/ℓ (0.05 and 0.1) and EI/GJ (1 and 5). Compare your results with those obtained from Eq. (3.92). Comment on the accuracy of the approximate solution in the text versus your modal solution. Which one should be more accurate? Discuss the trends of divergence dynamic pressure that you see regarding the sweep angle, stiffness ratio, and location of the aerodynamic center.
    (b) *Response*: For the response you will need to consider the nonhomogeneous equations, which should be put into the form

$$[[K] - \bar{q}[A]]\{\xi\} = \bar{q}\{\Xi_0\}.$$

Letting $\alpha_r = 1°$, obtain the response by solving the linear system of equations represented in this matrix equation. Plot the response of the wing tip (i.e., w and $\bar{\theta}$ at $\bar{y} = \ell$) for varying dynamic pressures up to $q = 0.95q_D$ for the above values of e/ℓ and EI/GJ with $\Lambda = -25°$ and $0°$. Plot the lift, twist, and bending moment distributions for the case with the largest tip twist angle. Comment on this result and on the trends of static aeroelastic response that you see regarding the sweep angle, stiffness ratio, and location of the aerodynamic center.

16. Consider the divergence of an unswept composite wing with $\kappa = 0$, $GJ/EI = 0.2$, and $e/\ell = .025$. Using Eq. (3.107), determine the value of κ, as defined by Eq. (3.106), needed to keep the divergence dynamic pressure unchanged for forward-swept wings with various values of $\Lambda < 0$. Plot these values of κ versus Λ.

17. Using the approximate formula found in Eq. (3.107), determine the divergence dynamic pressure for swept, composite wings when $e < 0$. Discuss the situations in which one might encounter a negative value of e. What sign of κ would you expect to be stabilizing in this case? Plot the divergence dynamic pressure for a swept composite wing with $GJ/EI = 0.2$ and $e/\ell = -.025$ versus $\kappa = 0$ and ± 0.4 for varying Λ.

18. Consider the divergence of a swept composite wing. Show that the governing equation and boundary conditions found in Eqs. (3.101) and (3.102) can be written as a second-order, integro-differential equation of the form

$$\theta'' + \tau\theta - r\tau \int_\eta^1 \theta(\xi)\,d\xi = 0,$$

with boundary conditions $\theta(0) = \theta'(1) = 0$ and with $r = \beta/\tau$. Determine the two simplest polynomial comparison functions for this reduced-order equation and boundary conditions. Use Galerkin's method to obtain one- and two-term approximations to the divergence dynamic pressure τ_D versus r. Plot your approximate solutions for the case in which $GJ/EI = 0.2$, $e/\ell = 0.02$, and $\kappa = -0.4$, depicted in Fig. 3.24, and compare these with the approximate solution given in the text. For the two-term approximation, determine the limit point for positive e, noting that the exact values are $r = 1.59768$ and $\tau_D = 10.7090$.

Answers: The one-term approximation is

$$\tau_D = \frac{30}{12 - 5r}.$$

The two-term approximation is

$$\tau_D = \frac{1260}{282 - 105r \pm \sqrt{3}\sqrt{15r(197r - 1{,}036) + 17{,}408}}.$$

The approximate limit point in the first quadrant is at $r = 1.61804$ and $\tau_D = 11.2394$. Within plotting accuracy, the two-term approximation is virtually indistinguishable from the exact solution when $-10 \leq \tau_D \leq 10$.

CHAPTER 4

Aeroelastic Flutter

The pilot of the airplane . . . succeeded in landing with roughly two-thirds of his horizontal tail surface out of action; some others have, unfortunately, not been so lucky. . . . The flutter problem is now generally accepted as a problem of primary concern in the design of current aircraft structures. Stiffness criteria based on flutter requirements are, in many instances, the critical design criteria. . . . There is no evidence that flutter will have any less influence on the design of aerodynamically controlled booster vehicles and re-entry gliders than it has, for instance, on manned bombers.

R. L. Bisplinghoff and H. Ashley in *Principles of Aeroelasticity* (John Wiley and Sons, Inc., 1962.)

Chapter 2 dealt with the subject of structural dynamics, which is the study of phenomena associated with the interaction of inertial and elastic forces in mechanical systems. In particular, the mechanical systems considered were one-dimensional continuous configurations that exhibit the general structural dynamic behavior of flight vehicles. If in the analysis of these structural dynamic systems aerodynamic loading is included, then the resulting dynamic phenomena may be classified as aeroelastic. As has been observed in Chapter 3, aeroelastic phenomena can have a significant influence on the design of flight vehicles. Indeed, these effects can greatly alter the design requirements that are specified for the disciplines of performance, structural loads, flight stability and control, and even propulsion. In addition, aeroelastic phenomena can introduce catastrophic instabilities of the structure that are unique to aeroelastic interactions and can limit the flight envelope.

Recalling the diagram in Fig. 1.1, one can classify aeroelastic phenomena as either static or dynamic. Whereas Chapter 3 dealt only with static aeroelasticity, in the present chapter we examine dynamic aeroelasticity. Although there are many other dynamic aeroelastic phenomena that could be treated, we focus in this chapter entirely on the instability called flutter, which generally leads to a catastrophic structural failure of the flight vehicle. A formal definition of aeroelastic flutter may be given as: *A dynamic instability of a flight vehicle associated with the interaction of aerodynamic, elastic, and inertial forces.* From this definition it is apparent that any investigation of flutter stability requires an adequate knowledge of the system's structural dynamic and aerodynamic properties.

Of the various phenomena that are categorized as aeroelastic flutter, lifting surface flutter is the one that is most often encountered and most likely to result in a catastrophic structural failure. As a result, it is required that all flight vehicle lifting surfaces be analyzed and tested to assure that this dynamic instability will not occur for any condition within the vehicle's flight envelope. If the airflow about the surface becomes separated during any portion of the elastic oscillation, the instability is called stall flutter and the governing equations become nonlinear. This type of instability most commonly occurs on turbojet compressor and helicopter rotor blades. Other phenomena that result in nonlinear behavior include large deflections, mechanical slop, and nonlinear control systems. Nonlinear phenomena will not be considered in the present treatment.

114

Even with this obvious paring down of the problem, one still finds that linear flutter analysis of clean lifting surfaces is complicated. We can only offer a simplified discussion of the theory of flutter. The reader is urged to consult the bibliography for additional reading on the subject.

This chapter begins by using the modal representation to set up a lifting surface flutter analysis as a linear set of ordinary differential equations. These are transformed into an eigenvalue problem, and the stability characteristics are then discussed in terms of the eigenvalues. Then, as an example of this methodology, a two-degree-of-freedom "typical section" analysis is formulated using the simple steady-flow aerodynamic model used in Chapter 3. The main shortcoming of this simple analysis is the neglect of unsteady effects in the aerodynamic model. Motivated by the need to consider unsteady aerodynamics in a meaningful but simple way, we then introduce classical flutter analysis. Engineering solutions that partially overcome the shortcomings of classical flutter analysis are then presented. To complete the set of analytical tools needed for flutter analysis, two very different unsteady aerodynamic theories are outlined, one suitable for use with classical flutter analysis and its derivatives, and the other suitable for eigenvalue-based flutter analysis. After illustrating how to approach the flutter analysis of a flexible wing using the assumed modes method, the chapter concludes with a discussion of flutter boundary characteristics.

4.1   Stability Characteristics

The lifting surface flutter of immediate concern can be described by a linear set of structural dynamic equations that include a linear representation of the unsteady airloads in terms of the elastic deformations. The surface could correspond to a wing or stabilizer either with or without control surfaces. Analytical simulation of the surface is sometimes made more difficult by the presence of external stores, engine nacelles, landing gear, or internal fuel tanks. Although such complexities complicate the analysis, they do not significantly alter the physical character of the flutter instability. For this reason the following discussion will be limited to a "clean" lifting surface.

When idealized for linear analysis, the nature of flutter is such that the flow over the lifting surface not only creates steady components of lift and pitching moment but also creates dynamic forces in response to small perturbations of the lifting surface motion, pitch and plunge motions in particular. Recall that the pitching motion of an airfoil may arise from torsional deformation, and the plunging of the airfoil may arise from bending deformation. When a lifting surface that is statically stable below its flutter speed is disturbed, the oscillatory motions caused by those disturbances will die out in time with exponentially decreasing amplitudes. That is, one could say that the air is providing damping for all such motions. Above the flutter speed, however, rather than damping out the motions caused by small perturbations in the configuration, the air can be said to be providing negative damping. Thus, these oscillatory motions grow with exponentially increasing amplitudes. This qualitative description of flutter can be observed in a general discussion of stability characteristics.

Before attempting to conduct an analysis of flutter, it is instructive to first examine the possible solutions to a structural dynamic representation in the presence of airloads. We will presume that the flight vehicle can be represented in terms of its normal modes of vibration. We illustrate this with the lifting surface modeled as a plate rather than a beam. This is somewhat more realistic for low-aspect-ratio wings, but in the present framework this increased realism presents very little increase in complexity because of the modal representation. For displacements $w(x, y, t)$ in the z direction normal to the plane of the planform (the x–y plane), the normal mode shapes can be represented by $\phi_i(x, y)$ and the associated natural frequencies by ω_i. A typical displacement of the structure can be written as

$$w(x, y, t) = \sum_{i=0}^{n} \xi_i(t)\phi_i(x, y), \tag{4.1}$$

where $\xi_i(t)$ is the generalized coordinate of the ith mode. It should be noted that the summation includes the index value $i = 0$, which symbolically denotes that rigid-body degrees of freedom have been included. The set of generalized equations of motion for the flight vehicle can be written as

$$M_i\left(\ddot{\xi}_i + \omega_i^2 \xi_i\right) = \Xi_i \qquad (i = 0, 1, \ldots, n), \tag{4.2}$$

where M_i is the generalized mass associated with the mass distribution, $m(x, y)$, and can be determined as

$$M_i = \iint\limits_{\text{planform}} m(x, y)\phi_i^2(x, y)\, dx\, dy. \tag{4.3}$$

The generalized force, $\Xi_i(t)$, associated with the external loading, $F(x, y, t)$, can be evaluated as

$$\Xi_i(t) = \iint\limits_{\text{planform}} F(x, y, t)\phi_i(x, y)\, dx\, dy. \tag{4.4}$$

To examine the stability properties of the flight vehicle, the only external loading to be considered is from the aerodynamic forces, which can be represented as a linear function of $w(x, y, t)$ and its time derivatives. It will be presumed that all other external disturbances have been eliminated. Such external disturbances would normally include atmospheric gusts, store ejection reactions, etc. Recalling that the displacement can be represented as a summation of the modal contributions, the induced pressure distribution, $\Delta p(x, y, t)$, can be described as a linear function of all the generalized coordinates and their derivatives. Such a relationship can be written as

$$\Delta p(x, y, t) = \sum_{j=0}^{n} [a_j(x, y)\xi_j(t) + b_j(x, y)\dot{\xi}_j(t) + c_j(x, y)\ddot{\xi}_j(t)]. \tag{4.5}$$

The corresponding generalized force of the ith mode can now be determined from

$$
\begin{aligned}
\Xi_i(t) &= \iint\limits_{\text{planform}} \Delta p(x, y, t)\phi_i(x, y)\, dx\, dy \\[2mm]
&= \sum_{j=0}^{n} \xi_j(t) \iint\limits_{\text{planform}} a_j(x, y)\phi_i(x, y)\, dx\, dy \\[2mm]
&\quad + \sum_{j=0}^{n} \dot{\xi}_j(t) \iint\limits_{\text{planform}} b_j(x, y)\phi_i(x, y)\, dx\, dy \\[2mm]
&\quad + \sum_{j=0}^{n} \ddot{\xi}_j(t) \iint\limits_{\text{planform}} c_j(x, y)\phi_i(x, y)\, dx\, dy \\[2mm]
&= \rho_\infty \frac{U^2}{b^2} \sum_{j=0}^{n} \left(a_{ij}\xi_j + \frac{b}{U}b_{ij}\dot{\xi}_j + \frac{b^2}{U^2}c_{ij}\ddot{\xi}_j \right).
\end{aligned}
\tag{4.6}
$$

Following the convention in some published work, we have factored out the freestream air density ρ_∞ and U^2/b^2 from the aerodynamic generalized force expression. Although not necessary, this step does enable the analyst to identify altitude effects more readily. It also shows explicitly that all aerodynamic effects vanish in a vacuum where ρ_∞ vanishes. Moreover, the normalization involving powers of b/U, where b is a reference semi-chord of the lifting surface, allows the matrices $[a]$, $[b]$, and $[c]$ to have the same units and is convenient for nondimensionalization later. Any nonhomogeneous terms in the generalized forces can be eliminated by redefinition of the generalized coordinates so that they are measured with respect to a different reference configuration. Thus, the generalized equations of motion can be written as a homogeneous set of differential equations when this form of the generalized force is included. They are

$$\frac{b^2}{U^2}(M_i\ddot{\xi}_i + M_i\omega_i^2\xi_i) - \rho_\infty\frac{b^2}{U^2}\sum_{j=0}^{n}c_{ij}\ddot{\xi}_j - \rho_\infty\frac{b}{U}\sum_{j=0}^{n}b_{ij}\dot{\xi}_j$$

$$-\rho_\infty\sum_{j=0}^{n}a_{ij}\xi_j = 0 \qquad (i = 0, 1, \ldots, n). \tag{4.7}$$

The general solution to this set of second-order, linear, ordinary differential equations can be described as a simple exponential function of time, because they are homogeneous. The form of this solution will be taken as

$$\xi_i(t) = \bar{\xi}_i\exp(\nu t). \tag{4.8}$$

Substitution of this expression into Eqs. (4.7) yields $n+1$ simultaneous linear, homogeneous, algebraic equations for the $\bar{\xi}_i$s since each term will contain an $\exp(\nu t)$. Thus,

$$\frac{b^2}{U^2}M_i\left(\nu^2 + \omega_i^2\right)\bar{\xi}_i - \rho_\infty\sum_{j=0}^{n}\left(\frac{b^2\nu^2}{U^2}c_{ij} + \frac{b\nu}{U}b_{ij} + a_{ij}\right)\bar{\xi}_j = 0 \quad (i = 0, 1, \ldots, n). \tag{4.9}$$

For a nontrivial solution of the generalized coordinate amplitudes, the determinant of the array formed by the coefficients of $\bar{\xi}_i$ must be zero. It is apparent that this determinant is a polynomial of degree $2(n+1)$ in ν. Subsequent solution of this polynomial equation for ν will typically yield $n+1$ complex conjugate pairs, represented as

$$\nu_k = \Gamma_k \pm i\Omega_k \qquad (k = 0, 1, \ldots, n). \tag{4.10}$$

For each ν_k there is a corresponding complex column matrix $\bar{\xi}_j^{(k)}$, $j = 0, 1, \ldots, n$. Thus, the solution of the generalized equations of motion with aerodynamic coupling can be written as

$$w(x, y, t) = \sum_{k=0}^{n}\{w_k(x, y)\exp\left[(\Gamma_k + i\Omega_k)t\right] + \bar{w}_k(x, y)\exp\left[(\Gamma_k - i\Omega_k)t\right]\}, \tag{4.11}$$

where $\bar{w}_k$ is the complex conjugate of w_k. This expression for $w(x, y, t)$ turns out to be real, as expected. Each w_k represents a unique linear combination of the mode shapes of the structure; that is,

$$w_k(x, y) = \sum_{i=0}^{n}\bar{\xi}_i^{(k)}\phi_i(x, y) \qquad (k = 0, 1, \ldots, n). \tag{4.12}$$

Note that only the relative values of $\bar{\xi}_i^{(k)}$ can be determined unless the initial displacement and rate of displacement are specified.

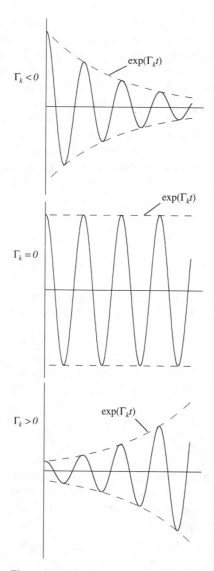

Figure 4.1   Behavior of typical mode amplitude when $\Omega_k \neq 0$.

It is apparent from the general solution for $w(x, y, t)$, Eq. (4.11), that the kth component of the summation represents a simple harmonic oscillation that is modified by an exponential function. The nature of this dynamic response to any specified initial condition is strongly dependent on the sign of each Γ_k. Typical response behavior is illustrated in Fig. 4.1 for positive, zero, and negative values of Γ_k when Ω_k is nonzero. We note that the negative of Γ_k is sometimes called the *modal damping* of the kth mode, and Ω_k is called the *modal frequency*. It is also possible to classify these motions from the standpoint of stability. The convergent oscillations when $\Gamma_k < 0$ are termed dynamically stable and the divergent oscillations for $\Gamma_k > 0$ are dynamically unstable. The case of $\Gamma_k = 0$ represents the boundary between the two and is often called the "stability boundary." If these solutions are for an aeroelastic system, the dynamically unstable condition is called *flutter*, and the stability boundary corresponding to simple harmonic motion is called the *flutter boundary*.

Table 4.1. *Types of Motion and Stability Characteristics for Various Values of* Γ_k *and* Ω_k

Γ_k	Ω_k	Type of Motion	Stability Characteristic
< 0	$\neq 0$	Convergent Oscillations	Stable
$= 0$	$\neq 0$	Simple Harmonic	Stability Boundary
> 0	$\neq 0$	Divergent Oscillations	Unstable
< 0	$= 0$	Continuous Convergence	Stable
$= 0$	$= 0$	Time Independent	Stability Boundary
> 0	$= 0$	Continuous Divergence	Unstable

Recall from Eq. (4.11) that the total displacement is a sum of all modal contributions. It is therefore necessary to consider all possible combinations of Γ_k and Ω_k, where Γ_k can be <0, $= 0$, or >0 and Ω_k can be $= 0$ or $\neq 0$. The corresponding type of motion and stability characteristics are indicated in Table 4.1 for various combinations of Γ_k and Ω_k. Although our primary concern here is with regard to the dynamic instability of flutter for which $\Omega_k \neq 0$, Table 4.1 shows that the generalized equations of motion can also provide solutions to the static aeroelastic problem of divergence. This phenomenon is indicated by the unstable condition for $\Omega_k = 0$, and the divergence boundary occurs when $\Gamma_k = \Omega_k = 0$.

In many published works on flutter analysis, the method outlined in this section is known as the p method, named for the reduced complex eigenvalue $p = bv/U$, which appears in Eq. (4.9) and in terms of which the eigenvalue problem can be posed instead of in terms of v as we have done. To provide accurate prediction of flutter characteristics the p method must use an aerodynamic theory that accurately represents the loads for transient motion of the lifting surface. Such theories may involve aerodynamic states, in addition to the structural generalized coordinates and their first time derivatives; see, for example, the theory outlined in Section 4.5.2. These additional states do not affect the meaning of the real and imaginary parts of the eigenvalues as discussed here. We now apply the p method to a simple configuration.

4.2   Aeroelastic Analysis of a Typical Section

In this section we will demonstrate the flutter analysis of a linear aeroelastic system. To do this a simple model is needed. In the older literature of aeroelasticity, flutter analyses were often performed using simple, spring-restrained, rigid-wing models such as the one shown in Fig. 4.2. These were called typical section models and are still very appealing

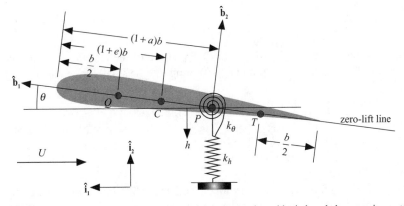

Figure 4.2   Schematic showing geometry of the wing section with pitch and plunge spring restraints.

because of their physical simplicity. This configuration could represent the case of a rigid, two-dimensional wind tunnel model that is elastically mounted in a wind tunnel test section, or it could correspond to a typical airfoil section along a finite wing. In the latter case the discrete springs would reflect the wing structural bending and torsional stiffnesses, and the reference point would represent the elastic axis.

Of interest in such models are points P, C, Q, and T, which refer, respectively, to the reference point (i.e., where the plunge displacement h is measured), the center of mass, the aerodynamic center (presumed to be the quarter-chord in thin-airfoil theory), and the three-quarter-chord (an important chordwise location in thin-airfoil theory). The dimensionless parameters e and a ($-1 \leq e \leq 1$ and $-1 \leq a \leq 1$) determine the locations of the points C and P; when these parameters are zero, the points lie on the mid-chord, and when they are positive (negative), the points lie toward the trailing (leading) edge. In the literature, the chordwise offset of the center of mass from the reference point, rather than e, often appears in the equations of motion. It is typically made dimensionless by the airfoil semichord b and denoted by $x_\theta = e - a$. This so-called static unbalance parameter is positive when the center of mass is toward the trailing edge from the reference point. The rigid plunging and pitching of the model is restrained by light, linear springs with spring constants k_h and k_θ.

It is convenient to formulate the equations of motion from Lagrange's equations. To do this, one needs kinetic and potential energies as well as the generalized forces resulting from aerodynamic loading. One can immediately write the potential energy as

$$P = \frac{1}{2}k_h h^2 + \frac{1}{2}k_\theta \theta^2. \tag{4.13}$$

To deduce the kinetic energy, one needs the velocity of the mass center C, which can be found as

$$\mathbf{v}_C = \mathbf{v}_P + \dot{\theta}\hat{\mathbf{b}}_3 \times b\left[(1 + a) - (1 + e)\right]\hat{\mathbf{b}}_1, \tag{4.14}$$

where the inertial velocity of the reference point P is

$$\mathbf{v}_P = -\dot{h}\hat{\mathbf{i}}_2, \tag{4.15}$$

and thus

$$\mathbf{v}_C = -\dot{h}\hat{\mathbf{i}}_2 + b\dot{\theta}(a - e)\hat{\mathbf{b}}_2. \tag{4.16}$$

The kinetic energy is then given by

$$K = \frac{1}{2}m\mathbf{v}_C \cdot \mathbf{v}_C + \frac{1}{2}I_C\dot{\theta}^2, \tag{4.17}$$

where I_C is the moment of inertia about C. By virtue of the relationship between $\hat{\mathbf{b}}_2$ and the inertially fixed unit vectors $\hat{\mathbf{i}}_1$ and $\hat{\mathbf{i}}_2$, assuming θ to be small, one finds that

$$\begin{aligned} K &= \frac{1}{2}m\left(\dot{h}^2 + b^2 x_\theta^2 \dot{\theta}^2 + 2bx_\theta \dot{h}\dot{\theta}\right) + \frac{1}{2}I_C\dot{\theta}^2 \\ &= \frac{1}{2}m(\dot{h}^2 + 2bx_\theta \dot{h}\dot{\theta}) + \frac{1}{2}I_P\dot{\theta}^2, \end{aligned} \tag{4.18}$$

where $I_P = I_C + mb^2 x_\theta^2$.

The generalized forces associated with the degrees of freedom h and θ are easily derived from the work done by the aerodynamic lift through a virtual displacement of the point Q

and by the aerodynamic pitching moment about Q through a virtual rotation of the model. The velocity of Q is

$$\mathbf{v}_Q = -\dot{h}\,\hat{\mathbf{i}}_2 + b\dot{\theta}\left(\frac{1}{2}+a\right)\hat{\mathbf{b}}_2. \tag{4.19}$$

The virtual displacement of the point Q can be obtained simply by replacing the dot over each unknown in Eq. (4.19) with a δ in front of it, that is,

$$\delta\mathbf{p}_Q = -\delta h\,\hat{\mathbf{i}}_2 + b\,\delta\theta\left(\frac{1}{2}+a\right)\hat{\mathbf{b}}_2, \tag{4.20}$$

where $\delta\mathbf{p}_Q$ is the virtual displacement at Q. The angular velocity of the wing is $\dot{\theta}\hat{\mathbf{b}}_3$, so that the virtual rotation of the wing is simply $\delta\theta\hat{\mathbf{b}}_3$. Therefore, the virtual work of the aerodynamic forces is

$$\overline{\delta W} = L\left[-\delta h + b\left(\frac{1}{2}+a\right)\delta\theta\right] + M_{\frac{1}{4}}\delta\theta \tag{4.21}$$

and the generalized forces become

$$Q_h = -L,$$
$$Q_\theta = M_{\frac{1}{4}} + b\left(\frac{1}{2}+a\right)L. \tag{4.22}$$

It is clear that the generalized force associated with h is the negative of the lift, whereas the one associated with θ is the pitching moment about the reference point P.

Lagrange's equations (as found in the Appendix, Eqs. A.35) are here specialized for the case in which the kinetic energy K depends only on $\dot{q}_1, \dot{q}_2, \ldots$, and so

$$\frac{d}{dt}\left(\frac{\partial K}{\partial\dot{q}_i}\right) + \frac{\partial P}{\partial q_i} = Q_i \qquad (i = 1, 2, \ldots, n). \tag{4.23}$$

Here $n = 2$, $q_1 = h$, and $q_2 = \theta$ and the equations of motion become

$$m(\ddot{h} + bx_\theta\ddot{\theta}) + k_h h = -L,$$
$$I_P\ddot{\theta} + mbx_\theta\ddot{h} + k_\theta\theta = M_{\frac{1}{4}} + b\left(\frac{1}{2}+a\right)L. \tag{4.24}$$

For the aerodynamics, the steady-flow theory we used in the previous chapter gives

$$L = 2\pi\rho_\infty bU^2\theta,$$
$$M_{\frac{1}{4}} = 0, \tag{4.25}$$

where, in accord with thin-airfoil theory, we have taken the lift-curve slope to be 2π. Assuming this representation to be adequate for the time being, we can apply the p method since the aerodynamic loads are specified for arbitrary motion. (We shall later consider more sophisticated aerodynamic theories.)

To simplify the notation, we introduce the uncoupled, natural frequencies at zero airspeed, defined by

$$\omega_h = \sqrt{\frac{k_h}{m}}, \qquad \omega_\theta = \sqrt{\frac{k_\theta}{I_P}}. \tag{4.26}$$

Substituting Eqs. (4.25) into Eqs. (4.24), making use of the definitions in Eqs. (4.26), and rearranging the equations of motion into matrix form, one obtains

$$
\begin{bmatrix} mb^2 & mb^2 x_\theta \\ mb^2 x_\theta & I_P \end{bmatrix} \begin{Bmatrix} \frac{\ddot{h}}{b} \\ \ddot{\theta} \end{Bmatrix}
$$

$$
+ \begin{bmatrix} mb^2 \omega_h^2 & 2\pi \rho_\infty b^2 U^2 \\ 0 & I_P \omega_\theta^2 - 2\left(\frac{1}{2}+a\right)\pi\rho_\infty b^2 U^2 \end{bmatrix} \begin{Bmatrix} \frac{h}{b} \\ \theta \end{Bmatrix} = \begin{Bmatrix} 0 \\ 0 \end{Bmatrix}. \tag{4.27}
$$

Note that the first equation has been multiplied through by b and the variable h has been divided by b to make every term in both equations have the same units. Following the p method as outlined above, we now make the substitutions $h = \overline{h}\exp(\nu t) = \overline{h}\exp(s\omega_\theta t)$ and $\theta = \overline{\theta}\exp(\nu t) = \overline{\theta}\exp(s\omega_\theta t)$, where s is an unknown, dimensionless, complex eigenvalue such that $s = \nu/\omega_\theta$, which gives

$$
\begin{bmatrix} mb^2 s^2 \omega_\theta^2 + mb^2 \omega_h^2 & mb^2 x_\theta s^2 \omega_\theta^2 + 2\pi\rho_\infty b^2 U^2 \\ mb^2 x_\theta s^2 \omega_\theta^2 & I_P s^2 \omega_\theta^2 + I_P \omega_\theta^2 - 2\left(\frac{1}{2}+a\right)\pi\rho_\infty b^2 U^2 \end{bmatrix} \begin{Bmatrix} \frac{\overline{h}}{b} \\ \overline{\theta} \end{Bmatrix} = \begin{Bmatrix} 0 \\ 0 \end{Bmatrix}.
$$
$$
\tag{4.28}
$$

Although this eigenvalue problem can be solved as it is written, it is very convenient to introduce dimensionless variables to further simplify the problem. To this end, we let

$$
r^2 = \frac{I_P}{mb^2}, \qquad \sigma = \frac{\omega_h}{\omega_\theta},
$$
$$
\mu = \frac{m}{\rho_\infty \pi b^2} \qquad V = \frac{U}{b\omega_\theta}. \tag{4.29}
$$

Here r is the dimensionless radius of gyration of the wing about the reference point P with $r^2 > x_\theta^2$, σ is the ratio of uncoupled bending to torsional frequencies, μ is the mass ratio parameter reflecting the relative importance of the model mass to the mass of the air affected by the model, and V is the dimensionless freestream speed of the air, sometimes called the reduced velocity. The equations then simplify to

$$
\begin{bmatrix} s^2 + \sigma^2 & s^2 x_\theta + \frac{2V^2}{\mu} \\ s^2 x_\theta & s^2 r^2 + r^2 - \frac{2V^2}{\mu}\left(\frac{1}{2}+a\right) \end{bmatrix} \begin{Bmatrix} \frac{\overline{h}}{b} \\ \overline{\theta} \end{Bmatrix} = \begin{Bmatrix} 0 \\ 0 \end{Bmatrix}. \tag{4.30}
$$

For a nontrivial solution to exist, the determinant of the coefficient matrix must be set equal to zero. There are two complex conjugate pairs of roots, say $s_1 = (\Gamma_1 \pm i\Omega_1)/\omega_\theta$ and $s_2 = (\Gamma_2 \pm i\Omega_2)/\omega_\theta$. For a given configuration and altitude one must look at the behavior of the complex roots as functions of V and find the smallest value of V to give divergent oscillations in accordance with Table 4.1. That value is $V_F = U_F/(b\omega_\theta)$, where U_F is the flutter speed.

It is noted that one may find the divergence speed by setting $s = 0$ in Eq. (4.30), which leads to setting the coefficient of $\overline{\theta}$ in the $\overline{\theta}$ equation equal to zero and solving the resulting expression for V. This value is the dimensionless divergence speed V_D, given by

$$
V_D = \frac{U_D}{b\omega_\theta} = r\sqrt{\frac{\mu}{1+2a}}. \tag{4.31}
$$

This is the same answer as one would obtain with an analysis similar to those of Chapter 3.

For looking at flutter, we consider a specific configuration defined by $a = -1/5$, $e = -1/10$, $\mu = 20$, $r^2 = 6/25$, and $\sigma = 2/5$. The divergence speed for this configuration is $V_D = 2.828$ (or $U_D = 2.828\, b\,\omega_\theta$). Plots of the imaginary and real parts of the roots versus

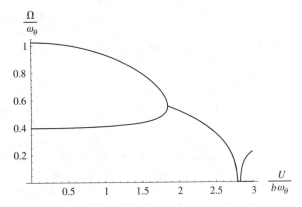

Figure 4.3 Plot of the modal frequency versus V for $a = -1/5$, $e = -1/10$, $\mu = 20$, $r^2 = 6/25$, and $\sigma = 2/5$ (steady-flow theory).

V are shown in Figs. 4.3 and 4.4, respectively. The negative of Γ is the modal damping, and Ω is the modal frequency. We consider first the imaginary parts, Ω, as shown in Fig. 4.3. When $V = 0$, one expects the two dimensionless frequencies to be near unity and σ for pitching and plunging oscillations, respectively. Even at $V = 0$ these modes are lightly coupled because of the nonzero off-diagonal term x_θ in the mass matrix. As V increases, the frequencies start to approach one another, and their respective mode shapes exhibit increasing coupling between plunge and pitch. Flutter occurs when the two modal frequencies coalesce, at which point the roots become complex conjugate pairs. At this condition, both modes are highly coupled pitch–plunge oscillations. The flutter speed is $V_F = U_F/(b\omega_\theta) = 1.843$, and the flutter frequency is $\Omega_F/\omega_\theta = 0.5568$. The real parts, Γ, are shown in Fig. 4.4 and remain zero until flutter occurs. When flutter occurs, the real part of one of the roots is positive and the other is negative.

Comparing results from the above analysis with experimental data, one finds that a few elements of realism are at least qualitatively captured. For example, the analysis predicts that flutter occurs at some value of $V = V_F < V_D$, which is correct for the specified configuration. Furthermore, it shows a coalescence of the pitching and plunging frequencies as

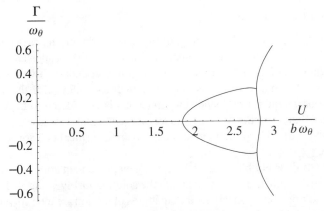

Figure 4.4 Plot of the modal damping versus V for $a = -1/5$, $e = -1/10$, $\mu = 20$, $r^2 = 6/25$, and $\sigma = 2/5$ (steady-flow theory).

V approaches V_F, which is not only correct for the specified configuration but is frequently observed in connection with flutter analysis. However, the above analysis is deficient in its ability to accurately predict the flutter speed. Moreover, the damping of all modes below the flutter speed is predicted to be zero, which is known to be incorrect.

The main reason for these deficiencies is that the aerodynamic theory from Chapter 3 was used. Although the aerodynamic theory has obvious deficiencies, such as its linearity and two-dimensionality, a very significant deficiency as far as flutter analysis is concerned is that it neglects unsteady effects, which are in general very important for flutter problems. The flow is unsteady because of two separate physical reasons. First, because of the wing's unsteady motion relative to the air, the relative wind vector is not fixed in space. Second, the airfoil motion disturbs the flow, shedding a vortex at the trailing edge. The downwash from this vortex, in turn, changes the flow that impinges on the airfoil. Thus, to obtain a more accurate prediction of the flutter speed, it is necessary to include unsteadiness in the aerodynamic theory. This demands a far more sophisticated aerodynamic theory.

Unfortunately, development of unsteady aerodynamic theories is no small undertaking. Unsteady aerodynamic theories can most simply be developed when simple harmonic motion is assumed *a priori*. Although such limited theories cannot be used in the p method of flutter analysis described in Section 4.1, they can be used in classical flutter analysis, described in the next section. As will be seen, classical flutter analysis can predict the flutter speed and flutter frequency, but it cannot predict values of modal damping and frequency away from the flutter condition. To obtain a reasonable sense of modal damping and frequencies at points other than the flutter condition, a couple of approximate schemes will be discussed in Section 4.4.

If these approximations turn out to be inadequate for predicting modal damping and frequencies, one has no choice but to carry out a flutter analysis that does not assume simple harmonic motion, which in turn requires a still more powerful aerodynamic theory. One such approach that fits easily into the framework of Section 4.1 is the finite-state theory of Peters et al. (1995). Such a theory not only facilitates the calculation of subcritical eigenvalues, but since it is a time-domain model it also can be used in control design.

Hence, in the sections to follow, we first look at classical flutter analysis and the approximate techniques associated therewith and then turn to a more detailed discussion of unsteady aerodynamics, including one theory that assumes simple harmonic motion (the Theodorsen theory) and one that does not (the Peters finite-state theory).

4.3   Classical Flutter Analysis

Throughout the aircraft industry most lifting surface flutter analyses performed are based on what is called a "classical flutter analysis." The objective of such an analysis is to determine the flight conditions that correspond to the flutter boundary. It was previously noted that the flutter boundary corresponds to conditions for which one of the modes of motion has a simple harmonic time dependency. Since this is considered to be a stability boundary, it is implied that all modes of motion are convergent (stable) for less critical flight conditions (lower airspeed). Moreover, all modes other than the critical one are convergent at the flutter boundary.

The method of analysis is not based on solving the generalized equations of motion as described in Section 4.1. Rather, it is presumed that the solution involves simple harmonic motion. With such a solution being specified, the equations of motion are then solved for the flight condition(s) that yield such a solution. Whereas in the p method one determines the eigenvalues for a set flight condition, the real parts of which provide the modal damping, it is apparent that classical flutter analysis cannot provide the modal damping for an arbitrary flight condition. Thus, it cannot provide any definitive measure of flutter stability other than the location of the stability boundary. Although this is the primary weakness of such a method, its primary strength is that it needs only the unsteady airloads for simple harmonic motion of the surface, which are more easily and thus more accurately obtained than those for arbitrary motion.

To illustrate classical flutter analysis it is necessary to consider an appropriate representation of unsteady airloads for simple harmonic motion of a lifting surface. Because these oscillatory motions are relatively small in amplitude, it is sufficient to use a linear aerodynamic theory for the computation of these loads. These aerodynamic theories are usually based on linear potential flow theory, which presumes that the motion of the structure is a small perturbation with respect to the freestream speed. For purposes of demonstration it will suffice to consider again the typical section of a two-dimensional lifting surface that is experiencing simultaneous translational and rotational motions, as illustrated in Fig. 4.2. The motion is simple harmonic; thus, h and θ will be represented as

$$
\begin{aligned}
h(t) &= \bar{h}\exp(i\omega t),\\
\theta(t) &= \bar{\theta}\exp(i\omega t),
\end{aligned}
\tag{4.32}
$$

where ω is the frequency of the motion. Although the h and θ motions are of the same frequency, they are not necessarily in phase. This can be taken into account mathematically by representing the amplitude $\bar{\theta}$ as a real number and $\bar{h}$ as a complex number. Since a linear aerodynamic theory is to be used the resulting lift, L, and the pitching moment about P, denoted by M where

$$
M = M_{\frac{1}{4}} + b\left(\frac{1}{2}+a\right)L,
\tag{4.33}
$$

will also be simple harmonic with frequency ω, so that

$$
\begin{aligned}
L(t) &= \bar{L}\exp(i\omega t),\\
M(t) &= \bar{M}\exp(i\omega t).
\end{aligned}
\tag{4.34}
$$

The amplitudes of these airloads can be computed as complex, linear functions of the amplitudes of motion as

$$
\begin{aligned}
\bar{L} &= -\pi\rho_\infty b^3\omega^2\left[\ell_h(k, M_\infty)\frac{\bar{h}}{b} + \ell_\theta(k, M_\infty)\bar{\theta}\right],\\
\bar{M} &= \pi\rho_\infty b^4\omega^2\left[m_h(k, M_\infty)\frac{\bar{h}}{b} + m_\theta(k, M_\infty)\bar{\theta}\right].
\end{aligned}
\tag{4.35}
$$

Here the freestream air density is represented as ρ_∞ and the four complex functions contained in the square brackets represent the dimensionless aerodynamic coefficients for the lift and moment resulting from plunging and pitching. These coefficients are in general functions of the two parameters k and M_∞, where

$$
\begin{aligned}
k &= \frac{b\omega}{U} \qquad \text{(reduced frequency)},\\
M_\infty &= \frac{U}{c_\infty} \qquad \text{(Mach number)}.
\end{aligned}
\tag{4.36}
$$

As in the case of steady airloads, compressibility effects are reflected here by the dependence of the coefficients on M_∞. The reduced frequency parameter k is unique to unsteady flows. This dimensionless frequency parameter is a measure of the unsteadiness of the flow and normally will have a value between zero and unity for conventional flight vehicles. It may also be noted that for any specified values of k and M_∞ each of the coefficients can be written as a complex number. As in the case of $\bar{h}$ relative to $\bar{\theta}$, the fact that lift and pitching moment are complex quantities reflects their phase relationships with respect to the pitch angle (where we can regard $\bar{\theta}$ as a real number for convenience). The speed at which flutter occurs corresponds to specific values of k and M_∞ and must be found by iteration. Examples of how this process can be carried out for one- and two-degree-of-freedom systems are given in this section.

4.3.1   *One-Degree-of-Freedom Flutter*

To illustrate the application of classical flutter analysis a very simple configuration will be treated first. This example is a one-degree-of-freedom aeroelastic system consisting of a rigid two-dimensional wing that is permitted to rotate in pitch about a specified reference point; this is a special case of the typical section configuration in Fig. 4.2 for which the plunge degree of freedom is equal to zero, as depicted in Fig. 4.5. The system equations of motion reduce to one equation that can be written as

$$I_P \ddot{\theta} + k_\theta \theta = M. \tag{4.37}$$

To be consistent with classical flutter analysis, the motion of the system will be presumed to be simple harmonic as

$$\theta = \bar{\theta} \exp(i\omega t). \tag{4.38}$$

The aerodynamic pitching moment, M, in the equation of motion is in response to this simple harmonic pitching displacement. As previously discussed this airload can be described by

$$M = \overline{M} \exp(i\omega t), \tag{4.39}$$

where

$$\overline{M} = \pi \rho_\infty b^4 \omega^2 m_\theta(k, M_\infty)\bar{\theta}. \tag{4.40}$$

Substituting these simple harmonic functions into the equation of motion yields an algebraic

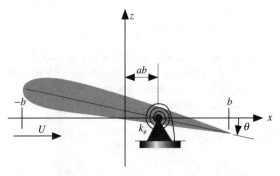

Figure 4.5   Schematic of the airfoil of a two-dimensional wing that is spring-restrained in pitch.

relation between the coefficients of $\overline{\theta}$ as

$$k_\theta - \omega^2 I_P = \pi \rho_\infty b^4 \omega^2 m_\theta(k, M_\infty). \tag{4.41}$$

Introducing the natural frequency of the system at zero airspeed,

$$\omega_\theta = \sqrt{\frac{k_\theta}{I_P}}, \tag{4.42}$$

and rearranging the algebraic relation, one obtains the final equation to be solved for the flight condition at the flutter boundary as

$$\frac{I_P}{\pi \rho_\infty b^4}\left[1 - \left(\frac{\omega_\theta}{\omega}\right)^2\right] + m_\theta(k, M_\infty) = 0. \tag{4.43}$$

To solve this equation it will be presumed that the configuration parameters I_P, ω_θ, and b are known. The unknown parameters that describe the motion and flight condition are ω, ρ_∞, k, and M_∞. These four unknowns must be determined from the single algebraic equation, Eq. (4.43). Since the aerodynamic coefficient, $m_\theta(k, M_\infty)$, is complex, it can be written as

$$m_\theta(k, M_\infty) = \Re(k, M_\infty) + i\Im(k, M_\infty). \tag{4.44}$$

As a consequence, both the real and imaginary parts of the algebraic relation must be zero, thus providing two real equations to determine the four unknowns. Therefore, two of the unknown parameters should be specified. A fixed altitude will be chosen that specifies the freestream atmospheric density, ρ_∞. The second parameter to be fixed will be the Mach number, which can be given a temporary value of zero. This, of course, implies that the flow is incompressible and the aerodynamic moment coefficient is then only a function of the reduced frequency. The governing algebraic equation can now be written as

$$\frac{I_P}{\pi \rho_\infty b^4}\left[1 - \left(\frac{\omega_\theta}{\omega}\right)^2\right] + \Re(k, 0) + i\Im(k, 0) = 0. \tag{4.45}$$

Equating the imaginary part of the left-hand side to zero gives a relation that can be solved for the reduced frequency, k_F, at the flutter boundary,

$$\Im(k_F, 0) = 0. \tag{4.46}$$

With k_F known $\Re(k_F, 0)$ can be numerically evaluated. Equating the real part of the left-hand side to zero now enables the frequency, ω_F, to be determined from

$$\left(\frac{\omega_\theta}{\omega_F}\right)^2 = 1 + \frac{\pi \rho_\infty b^4 \Re(k_F, 0)}{I_P}. \tag{4.47}$$

Now that k_F and ω_F have been determined, it is possible to compute the flutter speed as

$$U_F = \frac{b\omega_F}{k_F}. \tag{4.48}$$

The flutter speed determined by the above procedure corresponds to the originally specified altitude and is based on an incompressible representation of the airloads. After this speed has been determined, the speed of sound, c_∞, at the specified altitude can be used to find the flutter Mach number as

$$M_F = \frac{U_F}{c_\infty}. \tag{4.49}$$

If this flutter Mach number is sufficiently small to justify the use of incompressible aerodynamics coefficients, then the altitude–speed combination obtained is a point on the flutter boundary. If the flutter Mach number is too high to validate the incompressible approximation, then the entire procedure should be repeated using aerodynamic coefficients that are based on the initially computed flutter Mach number. Using the standard atmospheric model, which relates density and the speed of sound, this iterative scheme will converge to a flight condition on the flutter boundary.

4.3.2 *Two-Degree-of-Freedom Flutter*

The analysis of multi-degree-of-freedom systems for determination of the flutter boundary can be adequately demonstrated by the simple two-degree-of-freedom configuration of Fig. 4.2. The equations of motion, already derived as Eqs. (4.24), are

$$
\begin{aligned}
m(\ddot{h} + bx_\theta\ddot{\theta}) + k_h h &= -L, \\
I_P\ddot{\theta} + mbx_\theta\ddot{h} + k_\theta\theta &= M,
\end{aligned}
\tag{4.50}
$$

where, as before,

$$
M = M_{\frac{1}{4}} + b\left(\frac{1}{2} + a\right)L.
\tag{4.51}
$$

The next step in classical flutter analysis is to presume that the motion is simple harmonic as represented by

$$
\begin{aligned}
h &= \overline{h}\exp(i\omega t), \\
\theta &= \overline{\theta}\exp(i\omega t).
\end{aligned}
\tag{4.52}
$$

The corresponding lift and moment can be written as

$$
\begin{aligned}
L &= \overline{L}\exp(i\omega t), \\
M &= \overline{M}\exp(i\omega t).
\end{aligned}
\tag{4.53}
$$

Substituting these time-dependent functions into the equations of motion, one obtains a pair of algebraic equations for the amplitudes of h and θ of the form

$$
\begin{aligned}
-\omega^2 m\overline{h} - \omega^2 mbx_\theta\overline{\theta} + m\omega_h^2\overline{h} &= -\overline{L}, \\
-\omega^2 mbx_\theta\overline{h} - \omega^2 I_P\overline{\theta} + I_P\omega_\theta^2\overline{\theta} &= \overline{M},
\end{aligned}
\tag{4.54}
$$

where we recall that

$$
\begin{aligned}
\overline{L} &= -\pi\rho_\infty b^3\omega^2\left[\ell_h\left(k, M_\infty\right)\frac{\overline{h}}{b} + \ell_\theta\left(k, M_\infty\right)\overline{\theta}\right], \\
\overline{M} &= \pi\rho_\infty b^4\omega^2\left[m_h\left(k, M_\infty\right)\frac{\overline{h}}{b} + m_\theta\left(k, M_\infty\right)\overline{\theta}\right].
\end{aligned}
\tag{4.55}
$$

Substituting these lift and moment amplitudes into Eq. (4.54) and rearranging, one obtains a pair of homogeneous, linear, algebraic equations for $\bar{h}$ and $\bar{\theta}$, given by

$$\left\{\frac{m}{\pi\rho_\infty b^2}\left[1-\left(\frac{\omega_h}{\omega}\right)^2\right]+\ell_h\,(k,M_\infty)\right\}\frac{\bar{h}}{b}+\left[\frac{mx_\theta}{\pi\rho_\infty b^2}+\ell_\theta\,(k,M_\infty)\right]\bar{\theta}=0,$$

$$\left[\frac{mx_\theta}{\pi\rho_\infty b^2}+m_h\,(k,M_\infty)\right]\frac{\bar{h}}{b}+\left\{\frac{I_P}{\pi\rho_\infty b^4}\left[1-\left(\frac{\omega_\theta}{\omega}\right)^2\right]+m_\theta\,(k,M_\infty)\right\}\bar{\theta}=0.$$

$$(4.56)$$

The coefficients in these equations that involve the inertia terms will be symbolically simplified by defining the dimensionless parameters used earlier, namely,

$$\mu=\frac{m}{\pi\rho_\infty b^2}\qquad\text{(mass ratio)},$$

$$(4.57)$$

$$r=\sqrt{\frac{I_P}{mb^2}}\qquad\text{(mass radius of gyration about }P\text{)}.$$

Using these parameters allows us to rewrite the above two homogeneous equations in a somewhat simpler way:

$$\left\{\mu\left[1-\left(\frac{\omega_h}{\omega}\right)^2\right]+\ell_h\right\}\frac{\bar{h}}{b}+(\mu x_\theta+\ell_\theta)\bar{\theta}=0,$$

$$(4.58)$$

$$(\mu x_\theta+m_h)\frac{\bar{h}}{b}+\left\{\mu r^2\left[1-\left(\frac{\omega_\theta}{\omega}\right)^2\right]+m_\theta\right\}\bar{\theta}=0.$$

The third step in the flutter analysis is to solve these algebraic equations for the flight condition(s) for which the presumed simple harmonic motion is valid. This result will correspond to the flutter boundary. If it is presumed that the configuration parameters m, e, a, I_P, ω_h, ω_θ, and b are known, then the unknown quantities $\bar{h}$, $\bar{\theta}$, ω, ρ_∞, M_∞, and k describe the motion and flight condition. Because Eqs. (4.58) are linear and homogeneous in $\bar{h}/b$ and $\bar{\theta}$, the determinant of their coefficients must be zero for a nontrivial solution for the motion to exist. This condition can be written as

$$\left|\begin{array}{cc}\mu\left[1-\sigma^2\left(\frac{\omega_\theta}{\omega}\right)^2\right]+\ell_h & \mu x_\theta+\ell_\theta \\ \mu x_\theta+m_h & \mu r^2\left[1-\left(\frac{\omega_\theta}{\omega}\right)^2\right]+m_\theta\end{array}\right|=0.\qquad(4.59)$$

The determinant in this relation is called the "flutter determinant." It should be noted that the parameter $\sigma=\omega_h/\omega_\theta$ has been introduced, so that a common term explicit in ω is available, namely, ω_θ/ω. Thus, expansion of the determinant will yield a quadratic polynomial in the unknown $(\omega_\theta/\omega)^2$.

To complete the solution for the flight condition at the flutter boundary it must be recognized that four unknowns remain: ω_θ/ω, $\mu=m/(\pi\rho_\infty b^2)$, M_∞, and $k=b\omega/U$. The one equation available for their solution is the second-degree polynomial equation from the determinant. However, because the aerodynamic coefficients are complex quantities, this complex equation represents two real equations, wherein both the real and imaginary parts must be identically zero for a solution to be obtained. This means that two of the four unknowns must be specified. A procedure to solve for and map out the flutter boundary is outlined as follows:

1. Specify an altitude, which fixes the parameter μ.
2. Specify M_∞, say, to be zero.
3. Specify a set of trial k values, say from 0.001 to 1.0.
4. For each value of k (and the specified value of M_∞) calculate the functions ℓ_h, ℓ_θ, m_h, and m_θ.
5. Solve the flutter determinant, which is a quadratic equation with complex coefficients, for the values of $(\omega_\theta/\omega)^2$ that correspond to each of the selected values of k. Note that these roots will be complex in general, the real part being an approximation of $(\omega_\theta/\omega)^2$, and the imaginary part being related to the damping of the mode.
6. Interpolate to find the value of k at which the imaginary part of one of the roots becomes zero. This can be done approximately by plotting the imaginary parts of both roots versus k, so that the value of k at which one of the imaginary parts crosses the zero axis can be determined. This value of k then has a corresponding real value of $(\omega_\theta/\omega)^2$, which provides the value of ω.
7. Determine $U = b\omega/k$ and $M_\infty = U/c_\infty$.
8. Repeat steps 3–7 with the value of M_∞ obtained in step 7 until converged values are obtained for M_{∞_F}, k_F, and U_F for flutter at a given μ.
9. Repeat the whole procedure for various values of μ (an indication of the altitude for a given aircraft) to determine the flutter boundary in terms of, say, altitude versus M_{∞_F}, k_F, and U_F.

Step 6 above can also be carried out easily with computerized symbolic manipulation software such as Mathematica or Maple. One simply finds the value of k that makes the imaginary part of one of the two roots of the flutter determinant vanish.

4.4   Engineering Solutions for Flutter

It has been noted in the preceding section that the presumption of simple harmonic motion in classical flutter analysis has both advantages and disadvantages. The prime argument for specification of simple harmonic time dependency is, of course, its correspondence to the stability boundary. Identification of the flight conditions along this boundary requires the execution of a tedious, iterative process such as the one outlined above. This type of solution can be attributed to Theodorsen (1934), who presented the first comprehensive flutter analysis with his development of the unsteady airloads on a two-dimensional wing in an incompressible potential flow.

Although unsteady aerodynamics analyses for simple harmonic motion are not simple to formulate and execute, they are far more tractable than those for oscillatory motions with varying amplitude. Over the years since the work of Theodorsen, numerous unsteady aerodynamic formulations have been developed for simple harmonic motion of lifting surfaces. These techniques have proven to be quite adequate for compressible flows in both the subsonic and supersonic regimes. They have also been developed for three-dimensional surfaces and in some cases with surface-to-surface interaction. This availability of relatively accurate unsteady aerodynamic theories for simple harmonic motion has been the stimulus for further development of flutter analyses beyond that of the classical flutter analysis described in Section 4.3.

There are two other very important considerations of the practicing engineer. The first is to obtain an understanding of the margin of stability at flight conditions in the vicinity of the flutter boundary. The second, and possibly the more important, is to obtain an understanding of the physical mechanism that causes the instability. With these two pieces of information, the engineer can propose design variations that may alleviate or even eliminate the instability. When a suitable unsteady aerodynamic theory is available, the p method can address these considerations. In this section we look at alternative ways that engineers have addressed these problems when unsteady aerodynamic theories that assume simple harmonic motion must be used.

4.4.1 *The k Method*

Subsequent to Theodorsen's analysis of the flutter problem, numerous schemes were devised to extract the roots of the flutter determinant and thus identify the stability boundary. Scanlan and Rosenbaum (1951) presented a brief overview of these techniques as they were offered during the 1940s. It was fairly common to include in the flutter analysis a parameter that simulated the effect of structural damping. Observations at that time indicated that the energy removed per cycle during a simple harmonic oscillation was nearly proportional to the square of the amplitude but independent of the frequency. This behavior can be characterized by a damping force that is proportional to the displacement but in phase with the velocity.

To incorporate this form of structural damping into the analysis of Section 4.3.2, Eqs. (4.50) can be written as

$$m\left(\ddot{h} + bx_\theta\ddot{\theta}\right) + k_h h = -L + D_h,$$
$$I_P\ddot{\theta} + mbx_\theta\ddot{h} + k_\theta\theta = M + D_\theta,$$

(4.60)

where the dissipative structural damping terms are

$$D_h = \overline{D}_h \exp(i\omega t)$$
$$\quad = -ig_h m\omega_h^2 \overline{h} \exp(i\omega t),$$
$$D_\theta = \overline{D}_\theta \exp(i\omega t)$$
$$\quad = -ig_\theta I_p\omega_\theta^2\overline{\theta} \exp(i\omega t).$$

(4.61)

Proceeding as before, Eqs. (4.58) become

$$\left\{\mu\left[1 - \left(\frac{\omega_h}{\omega}\right)^2(1 + ig_h)\right] + \ell_h\right\}\frac{\overline{h}}{b} + (\mu x_\theta + \ell_\theta)\overline{\theta} = 0,$$
$$(\mu x_\theta + m_h)\frac{\overline{h}}{b} + \left\{\mu r^2\left[1 - \left(\frac{\omega_\theta}{\omega}\right)^2(1 + ig_\theta)\right] + m_\theta\right\}\overline{\theta} = 0.$$

(4.62)

The damping coefficients g_h and g_θ have representative values from 0.01 to 0.05 depending on the structural configuration. Most early analysts who incorporated this type of structural damping model into their flutter analyses specified the coefficient values *a priori* with the intention of improving the accuracy of their results.

It was Scanlan and Rosenbaum (1948) who suggested that the damping coefficients be treated as unknown together with ω. In this instance the subscripts on g can be removed. Writing $\sigma = \omega_h/\omega_\theta$ as before, and introducing

$$Z = \left(\frac{\omega_\theta}{\omega}\right)^2(1 + ig),$$

(4.63)

one obtains the flutter determinant as

$$\begin{vmatrix} \mu(1 - \sigma^2 Z) + \ell_h & \mu x_\theta + \ell_\theta \\ \mu x_\theta + m_h & \mu r^2 (1 - Z) + m_\theta \end{vmatrix} = 0, \tag{4.64}$$

which is a quadratic equation in Z. The two unknowns of this quadratic equation are complex, denoted by

$$Z_{1,2} = \left(\frac{\omega_\theta}{\omega_{1,2}} \right)^2 (1 + i g_{1,2}). \tag{4.65}$$

The computational strategy for solving Eq. (4.64) proceeds in a manner similar to the one outlined for Eq. (4.59). The primary difference is that the numerical results consist of two pairs of real numbers, (ω_1, g_1) and (ω_2, g_2), which can be plotted versus airspeed as either $U/(b\omega_\theta)$ or "reduced velocity" $1/k$.

Plots of the damping coefficient versus airspeed can indicate the margin of stability at conditions near the flutter boundary, where $g = 0$. These plots proved to be of such significance that the technique of incorporating the unknown structural damping was initially called the "U–g method," which is more indicative that it presumes simple harmonic motion throughout. The numerical values of g_i that are obtained for each k can only be interpreted as the required damping (of the specified form) to achieve simple harmonic motion at frequency ω_i. Because this damping does not really exist and has been introduced as an artifice to produce the desired motion, it is truly artificial structural damping.

The plots of frequency versus airspeed in conjunction with the damping plots can in many cases provide an indication of the physical mechanism that leads to the instability. The values of frequency along the $U = 0$ axis correspond to the coupled modes of the original structural dynamic system. As the airspeed increases the individual behavior or interaction of these roots can indicate the transfer of energy from one mode to another. Such observations could suggest a way to delay the onset of the instability. To confirm identification of the modes of motion for any specified reduced frequency, it is only necessary to substitute the corresponding eigenvalues, ω_i and g_i, back into the homogeneous equations of motion to compute the associated eigenvector $(\overline{h}/b, \overline{\theta})$. Since this is a complex number, it can provide the magnitude and phase of the original deflections h and θ.

4.4.2   The p–k Method

Although the k method provides significant advantages to the practicing aeroelastician, it is a mathematically improper formulation. The impropriety of imposing simple harmonic motion with the introduction of artificial damping has precipitated many heated discussions throughout the industry. It has been argued that for conditions other than the $g = 0$ case the frequency and damping characteristics do not correctly represent the system behavior. As a result design changes that are based on these characteristics can lead to expensive and potentially dangerous results.

In 1971 Hassig presented definitive numerical results that clearly indicated that the k method of flutter analysis can exhibit an improper coupling among the modes of motion. His presentation utilized a simple form of unsteady aerodynamics in a k method analysis, and he then compared the results with those from a p method analysis.

To illustrate the p method, the general solution to the homogeneous modal equations of motion given by Eqs. (4.8) can be written in terms of a dimensionless time, $\tau = Ut/b$, as

$$\xi_i(\tau) = \overline{\xi}_i \exp(p\tau). \tag{4.66}$$

Substitution of this expression into Eqs. (4.7) yields $n + 1$ linear, homogeneous equations for the $\bar{\xi}_i$s, which can be written in matrix form as

$$\left[p^2 [M] + \frac{b^2}{U^2} [M][\omega^2] - \rho_\infty [A(p)] \right] \{ \bar{\xi} \} = 0, \tag{4.67}$$

where it is noted that $[M]$ and $[\omega^2]$ are diagonal matrices with elements $M_1, M_2, \ldots, M_n$ and $\omega_1^2, \omega_2^2, \ldots, \omega_n^2$, respectively; n is the number of degrees of freedom; and the unsteady aerodynamic matrix is expressed as

$$[A(p)] = p^2 [c] + p [b] + [a], \tag{4.68}$$

where the matrices $[a]$, $[b]$, and $[c]$ are the same as those found in Eq. (4.9). Again, for a non-trivial solution of the generalized coordinate amplitudes, the determinant of the coefficient matrix in Eq. (4.67) must be zero, so that

$$\left| p^2 [M] + \frac{b^2}{U^2} [M][\omega^2] - \rho_\infty [A(p)] \right| = 0. \tag{4.69}$$

For a given speed and altitude this flutter determinant can be solved for p. The result will typically yield $n + 1$ complex conjugate pairs, represented as

$$p = \gamma k \pm i k, \tag{4.70}$$

where k is the reduced frequency of Eqs. (4.36), and γ is the rate of decay, given by

$$\gamma = \frac{1}{2\pi} \ln \left(\frac{a_{n+1}}{a_n} \right), \tag{4.71}$$

and where a_n and a_{n+1} represent the amplitudes of successive cycles.

Application of the k method to this modal representation can be readily achieved by letting $p = ik$ in the preceding formulation. This yields a flutter determinant comparable to Eq. (4.69) as

$$\left| -k^2 [M] + \frac{b^2}{U^2} [M][\omega^2] - \rho_\infty [A(ik)] \right| = 0. \tag{4.72}$$

At selected values of reduced frequency and altitude, Eq. (4.72) can be solved for the complex roots of b^2/U^2, denoted by $\lambda_r + i\lambda_i$. These roots may be interpreted as

$$\lambda_r + i\lambda_i = \left(\frac{b^2}{U^2} \right) (1 + ig), \tag{4.73}$$

where g is the structural damping required to sustain simple harmonic motion. This structural damping parameter can be related to the rate of decay parameter of the p method as

$$g \cong 2\gamma. \tag{4.74}$$

This is a good approximation for small damping as in the case of flight vehicles.

Another important aspect in making any correlation between the p and k methods is the matter of adequate inclusion of compressibility effects in the unsteady aerodynamic terms. In the p method the flutter determinant is solved for selected combinations of speed and altitude. Consequently, the appropriate Mach number can be used for the aerodynamic terms at the outset of the computation. In contrast, the k method preselects combinations of reduced frequency and altitude. As a result of then computing the airspeed as an unknown, λ_r, the Mach number cannot be accurately specified. The result is that an iterative process similar to the one in Section 4.3 must be conducted to ensure that compressibility effects are adequately incorporated in the k method.

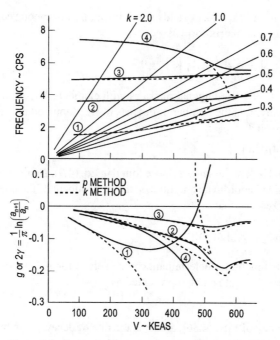

Figure 4.6 Comparison between p and k methods of flutter analysis for a twin-jet transport airplane. From Hassig (1971) Fig. 1, used by permission.

Hassig applied the p and k methods of flutter analysis to a realistic aircraft configuration. By incorporating the same unsteady aerodynamic representation in each analysis he was able to make a valid comparison of the results. His observations are typified by Fig. 4.6 (which is his Fig. 1). It can be noted from this figure that not only is the modal coupling wrongly predicted by the k method, but, more important, the wrong mode is predicted to become unstable. The only consistently valid result between the two analyses is that of the flutter speed for which $g = \gamma = 0$. In spite of the inconsistent modal coupling exhibited by the k method, it does permit the use of simple harmonic modeling of the unsteady aerodynamic terms. As previously mentioned the accuracy of simple harmonic airload predictions far exceeds the accuracy of airload predictions for transient motions. It is for this reason that a compromise between the two models has been suggested.

The p–k method is such a compromise. It is based on conducting a p method type of analysis with the restriction that the unsteady aerodynamics matrix is for simple harmonic motion. Using an estimated value of k in computing $[A(ik)]$, one finds the flutter determinant to be

$$\left| p^2[M] + \frac{b^2}{U^2}[M][\omega^2] - \rho_\infty[A(ik)] \right| = 0. \tag{4.75}$$

Given a set of initial guesses for k, say $k_0 = b\omega_i/U$ for the ith root, this equation can be solved for p. This typically yields a set of complex conjugate pairs of roots, the number of which corresponds to the number of degrees of freedom in the structural model. For one of the roots, denoting the initial solution as

$$k_1 = |\Im(p)|, \qquad \gamma_1 = \frac{\Re(p)}{k_1}, \tag{4.76}$$

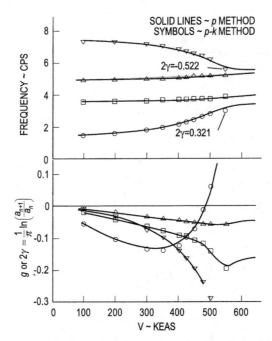

Figure 4.7 Comparison between p and $p{-}k$ methods of flutter analysis for a twin-jet transport airplane. From Hassig (1971) Fig. 2, used by permission.

one can compute $[A(ik_1)]$. Using this new matrix in Eq. (4.75) leads to another set of ps, so that

$$k_2 = |\Im(p)|, \qquad \gamma_2 = \frac{\Re(p)}{k_2}. \tag{4.77}$$

Continual updating of the aerodynamic matrix in this way provides an iterative scheme that is convergent for each of the roots, with negative γ being a measure of the modal damping. The earliest presentation of this technique was offered by Irwin and Guyett in 1965.

Hassig applied the $p{-}k$ method to the configuration of Fig. 4.6. As illustrated by Fig. 4.7 (which is his Fig. 2), the $p{-}k$ method appears to yield approximately the same result as the p method. This, of course, simply validates the convergence of the scheme. Its greatest advantage is that it can utilize airloads that have been formulated for simple harmonic motion. Another comparison offered by Hassig was between the widely used k method and the $p{-}k$ method for a horizontal stabilizer/elevator configuration. This example of a strongly coupled system provided the results given in Fig. 4.8 (which is his Fig. 3). Here again, as in the k versus p comparison of Fig. 4.6, widely differing conclusions can be drawn regarding the modal coupling. In addition to the easily interpreted frequency and damping plots versus airspeed for strongly coupled systems, a second advantage is offered by the $p{-}k$ method regarding computational effort. The k method requires numerous computer runs at constant density to ensure matching the Mach number with airspeed and altitude. The $p{-}k$ method does not have this requirement.

The accuracy of the $p{-}k$ method depends on the level of damping in any particular mode. It is left as an exercise to the reader (see Problem 13) to show that the $p{-}k$ method damping is only a good approximation for the damping in lightly damped modes. Fortunately, these are the modes about which we care the most.

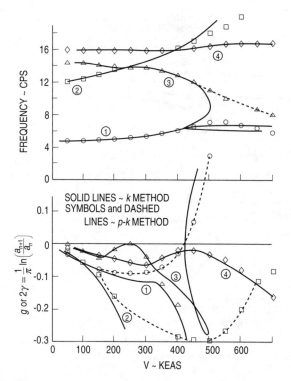

Figure 4.8   Comparison between p–k and k methods of flutter analysis for a horizontal stabilizer with elevator. From Hassig (1971) Fig. 3, used by permission.

4.5   Unsteady Aerodynamics

In Section 4.2, flutter analysis was conducted using an aerodynamic theory for steady flow. The lift and pitching moment used were functions only of the instantaneous pitch angle, θ. Fung (1955) suggested a simple experiment to demonstrate that things are not that simple: Attempt to rapidly move a stick in a straight line through water and notice the results. In the wake of the stick there is a vortex pattern, with vortices being shed alternately from each side of the stick. This shedding of vortices induces a periodic force perpendicular to the stick's line of motion, causing the stick to tend to wobble back and forth in your hand. A similar phenomenon happens with the motion of a lifting surface through a fluid and must be accounted for in unsteady aerodynamic theories.

In a more complete unsteady aerodynamic theory, the lift and pitching moment consist of two parts from two physically different phenomena: noncirculatory and circulatory effects. Noncirculatory effects, also called apparent mass and inertia effects, are generated when the wing motion has a nonzero acceleration. It has to then carry with it a part of the air surrounding it. That air has finite mass, which leads to inertial forces opposing its acceleration.

Circulatory effects are generally more important for aircraft wings. Indeed, in steady flight it is the circulatory lift that keeps the aircraft aloft. Vortices are an integral part of the process of generation of circulatory lift. Basically, there is a difference in the velocities on the upper and lower surfaces of an airfoil. Such a velocity profile can be represented as a constant velocity flow plus a vortex. In a dynamic situation, the strength of the vortex (i.e., the circulation) is changing with time. However, the circulatory forces of steady-flow theories do not include the effects of the vortices shed into the wake. Restricting our discussion to two dimensions and potential flow, we recall an implication of the Helmholtz theorem: The total vorticity will always vanish within any closed curve surrounding a particular set of fluid particles. Thus, if some clockwise vorticity develops about the airfoil, a counterclockwise vortex of the same strength has to be shed into the flow. As it moves along this shed vortex changes the flow field by inducing an unsteady flow back onto the airfoil. This behavior is a function of the strength of the shed vortex and its distance away from the airfoil. Thus, accounting for the effect of shed vorticity is in general a very complex undertaking and would necessitate knowledge of each and every vortex shed in the flow. However, if one assumes that the vortices shed in the flow move with the flow, then one can estimate the effect of these vortices.

In this section we present two types of unsteady aerodynamic theories, both of which are based on potential flow theory and take into account the effects of shed vorticity. The simpler theory is appropriate for classical flutter analysis as well as for the k and p–k methods. The other is a finite-state theory cast in the time domain, appropriate for time-domain analysis as well as for eigen-analysis in the form of the p method.

4.5.1   *Theodorsen's Unsteady Thin-Airfoil Theory*

Theodorsen (1934) derived a theory of unsteady aerodynamics for a thin airfoil undergoing small oscillations in incompressible flow. The lift contains both circulatory and noncirculatory terms, whereas the pitching moment about the quarter-chord is entirely noncirculatory. According to Theodorsen's theory, the lift and pitching moment are given by

$$L = 2\pi\rho_\infty U b C(k) \left[\dot{h} + U\theta + b\left(\frac{1}{2} - a\right)\dot{\theta} \right] + \pi\rho_\infty b^2(\ddot{h} + U\dot{\theta} - ba\ddot{\theta}),$$

$$M_{\frac{1}{4}} = -\pi\rho_\infty b^3 \left[\frac{1}{2}\ddot{h} + U\dot{\theta} + b\left(\frac{1}{8} - \frac{a}{2}\right)\ddot{\theta}\right], \tag{4.78}$$

where the generalized forces are given in Eqs. (4.22). The function $C(k)$ is a complex-valued function of the reduced frequency k, given by

$$C(k) = \frac{H_1^{(2)}(k)}{H_1^{(2)}(k) + i H_0^{(2)}(k)}, \tag{4.79}$$

where $H_n^{(2)}(k)$ are Hankel functions of the second kind, which can be expressed in terms of Bessel functions of the first and second kind, respectively, as

$$H_n^{(2)}(k) = J_n(k) - i Y_n(k). \tag{4.80}$$

The function $C(k)$ is called Theodorsen's function and is plotted in Fig. 4.9. Note that $C(k)$ is real and equal to unity for the steady case (i.e., for $k = 0$). As k increases, one finds that the imaginary part increases in magnitude while the real part decreases. As k tends to infinity, $C(k)$ approaches $1/2$. However, for practical situations k does not exceed values of the order of unity. Hence, the plot in Fig. 4.9 only extends to $k = 1$. When any harmonic function is multiplied by $C(k)$, its magnitude is reduced and a phase lag is introduced.

A few things are noteworthy concerning Eqs. (4.78). First, in Theodorsen's theory the lift-curve slope is equal to 2π. Thus, the first of the two terms in the lift is the circulatory

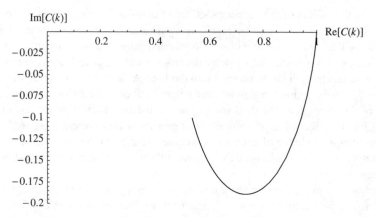

Figure 4.9 Plot of the real and imaginary parts of $C(k)$ for k varying from zero, where $C(k) = 1$, to unity.

lift without the effect of shed vortices multiplied by $C(k)$. The multiplication by $C(k)$ is a consequence of the theory having taken into account the effect of shed vorticity. The second term in the lift as well as the pitching moment are noncirculatory, depending on the acceleration and angular acceleration of the airfoil. The circulatory lift is the more significant of the two terms in the lift.

For steady flow the circulatory lift is linear in the angle of attack, but for unsteady flow there is no single angle of attack since the flow direction varies along the chord-line as the result of the induced flow. However, just so we can discuss the concept for unsteady flow, it is possible to introduce a so-called effective angle of attack. For simple harmonic motion it can be inferred from Theodorsen's theory that an effective angle of attack is

$$\alpha = C(k) \left[\theta + \frac{\dot{h}}{U} + \frac{b}{U} \left(\frac{1}{2} - a \right) \dot{\theta} \right]. \tag{4.81}$$

As we will show in Section 4.5.2 by comparison with the finite-state aerodynamic model introduced therein, α is the angle of attack measured at the three-quarter chord based on an averaged value of the induced flow. Recall that in steady-flow aerodynamics, the angle of attack is the pitch angle θ. Here, however, α depends on θ as well as on $\dot{h}, \dot{\theta}$, and k. Because of these additional terms and because of the behavior of $C(k)$, we expect changes in magnitude and phase between θ and α carrying over into changes in the magnitude and phase of the lift and pitching moment relative to that of θ. Indeed, the function $C(k)$ is sometimes called the lift-deficiency function because it reduces the magnitude of unsteady lift relative to steady lift. It also introduces an important phase shift between the peak values of pitching oscillations and corresponding oscillations in lift and pitching moment. An approximation of Theodorsen theory in which $C(k)$ is set equal to unity is called a "quasi-steady" thin-airfoil theory. Such an approximation has value only for cases in which k is restricted to be very small.

Theodorsen's theory may be used in classical flutter analysis. There the reduced frequency of flutter is not known *a priori*. One can find k at the flutter condition using the method described in Section 4.3. Theodorsen's theory may also be used in the k and p–k methods, as described in Sections 4.4.1 and 4.4.2, respectively.

4.5.2   *Finite-State Unsteady Thin-Airfoil Theory of Peters et al.*

Although the Theodorsen theory is an excellent choice for classical flutter analysis, there are situations in which an alternative approach is needed. First, frequently one needs to calculate the modal damping in subcritical flight conditions. Second, there is a growing interest in the active control of flutter, and design of controllers requires that the system be represented in state-space form. To meet these requirements, one needs to represent the actual aerodynamic loads (which are in the frequency domain in Theodorsen's theory) in terms of time-domain differential equations. Finite-state theories approximate the actual infinite-state aerodynamic model to within engineering accuracy. One such approach is the finite-state, induced-flow theory for inviscid, incompressible flow of Peters et al.

Consider a typical section of a rigid, symmetric wing, shown in Fig. 4.2, and the additional vectorial directions defined in Fig. 4.10. To begin the presentation of this theory, we first relate the three sets of unit vectors:

1. a set fixed in the inertial frame, $\hat{\mathbf{i}}_1$ and $\hat{\mathbf{i}}_2$, such that the air is flowing at velocity $-U\hat{\mathbf{i}}_1$,
2. a set fixed in the wing, $\hat{\mathbf{b}}_1$ and $\hat{\mathbf{b}}_2$, with $\hat{\mathbf{b}}_1$ directed along the zero-lift line toward the leading edge and $\hat{\mathbf{b}}_2$ perpendicular to $\hat{\mathbf{b}}_1$,
3. a set $\hat{\mathbf{a}}_1$ and $\hat{\mathbf{a}}_2$ associated with the local relative wind vector at the three-quarter chord, such that $\hat{\mathbf{a}}_1$ is along the relative wind vector and $\hat{\mathbf{a}}_2$ is perpendicular to it, in the assumed direction of the lift.

The relationships among these unit vectors can be simply stated as

$$\begin{Bmatrix} \hat{\mathbf{b}}_1 \\ \hat{\mathbf{b}}_2 \end{Bmatrix} = \begin{bmatrix} \cos(\theta) & \sin(\theta) \\ -\sin(\theta) & \cos(\theta) \end{bmatrix} \begin{Bmatrix} \hat{\mathbf{i}}_1 \\ \hat{\mathbf{i}}_2 \end{Bmatrix} \tag{4.82}$$

and

$$\begin{Bmatrix} \hat{\mathbf{a}}_1 \\ \hat{\mathbf{a}}_2 \end{Bmatrix} = \begin{bmatrix} \cos(\alpha) & -\sin(\alpha) \\ \sin(\alpha) & \cos(\alpha) \end{bmatrix} \begin{Bmatrix} \hat{\mathbf{b}}_1 \\ \hat{\mathbf{b}}_2 \end{Bmatrix} \tag{4.83}$$

and $\hat{\mathbf{i}}_3 = \hat{\mathbf{a}}_3 = \hat{\mathbf{b}}_3 = \hat{\mathbf{b}}_1 \times \hat{\mathbf{b}}_2$.

Induced-flow theories approximate the effects of shed vortices based on changes they cause in the flow field near the airfoil. Thus, the velocity field near the airfoil consists of the

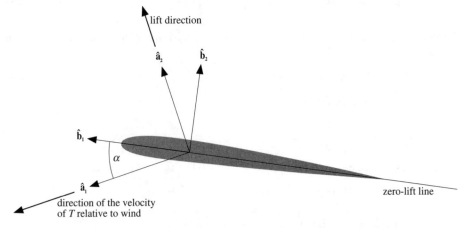

Figure 4.10   Schematic showing geometry of the zero-lift line, relative wind, and lift directions.

freestream velocity plus an additional component to account for the induced flow. Although the induced flow varies throughout the flow field, we will approximate its value near the airfoil as an average value along the chordline. Thus, the local inertial wind velocity is written approximately as $-U\hat{\mathbf{i}}_1 - \lambda_0\hat{\mathbf{b}}_2$, where λ_0 is the average induced flow (positive as downwash normal to the airfoil zero-lift line). According to classical thin-airfoil theory, one should calculate the angle of attack using the instantaneous relative wind velocity vector as calculated at T. To represent the relative wind velocity vector at T, one can write the relative wind vector (i.e., the velocity of the wing with respect to the air) as $W\hat{\mathbf{a}}_1$ and set it equal to the inertial velocity of T minus the inertial air velocity, that is,

$$W\hat{\mathbf{a}}_1 = \mathbf{v}_T - (-U\hat{\mathbf{i}}_1 - \lambda_0\hat{\mathbf{b}}_2)$$

$$= \mathbf{v}_T + U\hat{\mathbf{i}}_1 + \lambda_0\hat{\mathbf{b}}_2, \qquad (4.84)$$

where $\mathbf{v}_T$ is the inertial velocity of the three-quarter chord, given by

$$\mathbf{v}_T = \mathbf{v}_P + \dot{\theta}\hat{\mathbf{b}}_3 \times \mathbf{r}_{PT}, \qquad (4.85)$$

and $\mathbf{r}_{PT}$ is the position vector from P to T. From Fig. 4.2 one finds that

$$\mathbf{r}_{PT} = \left[\frac{b}{2} + (1+a)b - 2b\right]\hat{\mathbf{b}}_1 = b\left(a - \frac{1}{2}\right)\hat{\mathbf{b}}_1. \qquad (4.86)$$

Thus,

$$\mathbf{v}_T = \mathbf{v}_P + \dot{\theta}\hat{\mathbf{b}}_3 \times b\left(a - \frac{1}{2}\right)\hat{\mathbf{b}}_1. \qquad (4.87)$$

The inertial velocity of the reference point P is

$$\mathbf{v}_P = -\dot{h}\hat{\mathbf{i}}_2, \qquad (4.88)$$

while $\dot{\theta}\hat{\mathbf{b}}_3$ is the inertial angular velocity of the wing. Carrying out the cross product in Eq. (4.87), one obtains

$$\mathbf{v}_T = -\dot{h}\hat{\mathbf{i}}_2 + b\dot{\theta}\left(a - \frac{1}{2}\right)\hat{\mathbf{b}}_2, \qquad (4.89)$$

so that the relative wind can be written as

$$W\hat{\mathbf{a}}_1 = U\hat{\mathbf{i}}_1 - \dot{h}\hat{\mathbf{i}}_2 + \left[b\dot{\theta}\left(a - \frac{1}{2}\right) + \lambda_0\right]\hat{\mathbf{b}}_2. \qquad (4.90)$$

Alternatively, one may write the relative wind in terms of its components along $\hat{\mathbf{b}}_1$ and $\hat{\mathbf{b}}_2$, that is,

$$W\hat{\mathbf{a}}_1 = W\cos(\alpha)\hat{\mathbf{b}}_1 - W\sin(\alpha)\hat{\mathbf{b}}_2, \qquad (4.91)$$

where α is given by (see Fig. 4.10)

$$\tan(\alpha) = -\frac{\hat{\mathbf{a}}_1 \cdot \hat{\mathbf{b}}_2}{\hat{\mathbf{a}}_1 \cdot \hat{\mathbf{b}}_1}. \qquad (4.92)$$

Using Eq. (4.90), one finds that

$$W\hat{\mathbf{a}}_1 \cdot \hat{\mathbf{b}}_1 = U\cos(\theta) - \dot{h}\sin(\theta),$$

$$W\hat{\mathbf{a}}_1 \cdot \hat{\mathbf{b}}_2 = -U\sin(\theta) - \dot{h}\cos(\theta) + b\dot{\theta}\left(a - \frac{1}{2}\right) + \lambda_0. \qquad (4.93)$$

Assuming small angles, one may now show that

$$\alpha = \theta + \frac{\dot{h}}{U} + \frac{b}{U}\left(\frac{1}{2} - a\right)\dot{\theta} - \frac{\lambda_0}{U},$$

$$W = U + \text{higher-order terms.}$$

(4.94)

According to this derivation, α is an effective angle of attack based on the relative wind vector at the three-quarter chord, which, in turn, is based on the average value of the induced flow λ_0 over the wing chordline. Note that α is *not* equal to the pitch angle θ. Because of the motion of the wing and the induced flow field, the relative wind direction is not fixed in inertial space. Therefore, the effective angle of attack depends on the pitch rate, on the plunge velocity, and on the induced flow. Moreover, the lift is assumed to be perpendicular to the relative wind vector. This assumption is adequate for the calculation of lift and pitching moment, which are both first order in the motion variables. However, sufficiently rapid plunge motion (as in the flapping wings of an insect) can result in a value of α that is not small, and one would need to make "small but finite" angle assumptions to calculate the drag (or propulsive force equal to negative drag) one could encounter in such situations.

The total lift and moment expressions including the noncirculatory forces are

$$L = \pi\rho_\infty b^2(\ddot{h} + U\dot{\theta} - ba\ddot{\theta}) + 2\pi\rho_\infty Ub\left[\dot{h} + U\theta + b\left(\frac{1}{2} - a\right)\dot{\theta} - \lambda_0\right],$$

$$M_{\frac{1}{4}} = -\pi\rho_\infty b^3\left[\frac{1}{2}\ddot{h} + U\dot{\theta} + b\left(\frac{1}{8} - \frac{a}{2}\right)\ddot{\theta}\right].$$

(4.95)

Note the similarity between Eqs. (4.95) and (4.78). In particular, by studying the circulatory lift in both lift equations, one then can see the basis for identifying α, calculated as in the first of Eqs. (4.94), with the expression in Eq. (4.81).

The lift and pitching moment are then used to form the generalized forces from Eqs. (4.22) and are, in turn, used in the structural equations, Eqs. (4.24). Even so, these two equations are incomplete, having more than two unknowns. The induced flow velocity λ_0 needs to be expressed in terms of the airfoil motion. The induced-flow theory of Peters et al. does just that, representing the average induced flow λ_0 in terms of N induced-flow states $\lambda_1, \lambda_2, \dots,$ λ_N as

$$\lambda_0 \approx \frac{1}{2}\sum_{n=1}^{N} b_n\lambda_n,$$

(4.96)

where the b_n are found by the least-squares method. The induced-flow dynamics are then derived from the assumption that the shed vortices stay in the plane of the airfoil and travel downstream with the same velocity as the flow. Introducing a column matrix $\{\lambda\}$ containing the values of λ_n, one can write the set of N first-order ordinary differential equations governing $\{\lambda\}$ as

$$[A]\{\dot{\lambda}\} + \frac{U}{b}\{\lambda\} = \{c\}\left[\ddot{h} + U\dot{\theta} + b\left(\frac{1}{2} - a\right)\ddot{\theta}\right],$$

(4.97)

where the matrices $[A]$ and $\{c\}$ can be derived for a user-defined number of induced-flow states. The expressions of the matrices used above are given for N finite states as

$$[A] = [D] + \{d\}\{b\}^T + \{c\}\{d\}^T + \frac{1}{2}\{c\}\{b\}^T,$$

(4.98)

where

$$D_{nm} = \begin{cases} \frac{1}{2n}, & n = m + 1, \\ -\frac{1}{2n}, & n = m - 1, \\ 0, & n \neq m \pm 1, \end{cases} \tag{4.99}$$

$$b_n = \begin{cases} (-1)^{n-1} \frac{(N+n-1)!}{(N-n-1)!} \frac{1}{(n!)^2}, & n \neq N, \\ (-1)^{n-1}, & n = N, \end{cases} \tag{4.100}$$

$$d_n = \begin{cases} \frac{1}{2}, & n = 1, \\ 0, & n \neq 1, \end{cases} \tag{4.101}$$

and

$$c_n = \frac{2}{n}. \tag{4.102}$$

The resulting aeroelastic model is in the time domain, in contrast to classical flutter analysis, which is in the frequency domain (see Section 4.3). Thus, it can be used for flutter analysis by the p method as well as in the design of control systems to alleviate flutter.

Results using the finite-state, induced-flow model (i.e., Eqs. 4.97 and 4.24 with generalized forces given by Eqs. 4.22 with lift and pitching moment given by Eqs. 4.95) for the problem analyzed previously in Section 4.2 (recall that $a = -1/5$, $e = -1/10$, $\mu = 20$, $r^2 = 6/25$, and $\sigma = 2/5$) are given here. These results are based on use of $N = 6$ induced-flow states. The frequency and damping results are shown in Figs. 4.11 and 4.12, respectively. As before, a frequency coalescence is observed near where one of the modes goes unstable. The flutter speed obtained is $V_F = U_F/(b\omega_\theta) = 2.165$, and the flutter frequency is $\Omega_F/\omega_\theta = 0.6545$. Although this value of the flutter speed is close to that observed previously using the simpler theory, the unsteady theory produces complex roots for all $V \neq 0$, so that there is modal damping in all the modes below the flutter speed. The equations contain damping terms proportional to the velocity that account for the initial increase in damping. At higher velocities, however, the destabilizing circulatory term (the nonsymmetric term in

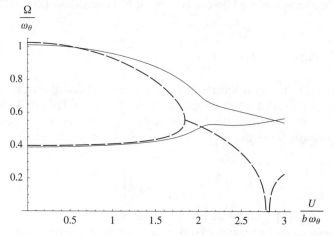

Figure 4.11 Plot of the modal frequency versus $U/(b\omega_\theta)$ for $a = -1/5$, $e = -1/10$, $\mu = 20$, $r^2 = 6/25$, and $\sigma = 2/5$; solid lines: p method, aerodynamics of Peters et al.; dashed lines: steady flow aerodynamics.

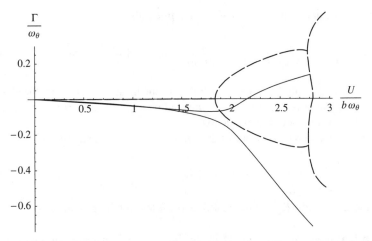

Figure 4.12   Plot of the modal damping versus $U/(b\omega_\theta)$ for $a = -1/5$, $e = -1/10$, $\mu = 20$, $r^2 = 6/25$, and $\sigma = 2/5$; solid lines: p method, aerodynamics of Peters et al.; dashed lines: steady flow aerodynamics.

the stiffness matrix that also is present in the steady-flow theory) overcomes the damping caused by the unsteady terms, resulting in flutter.

It is left to the reader as an exercise to show the equivalence of the theory of Peters et al. and Theodorsen's theory; see Problem 15.

4.6   Flutter Prediction via Assumed Modes

As previously noted, in industry it is now typical to use the finite element method as a means to realistically represent aircraft structural dynamics. Although it is certainly possible to conduct full finite element flutter analyses, flutter analysis based on a truncated set of the modes of the stucture is still very helpful and relatively simple; and for those reasons alone it is often done. In this section we show how such an analysis can be done within the framework of the Ritz method, already established in Section 2.4.

Let us consider an unswept wing that is modeled as a beam of length ℓ. For the structural model, we adopt the same notation as we used in Chapter 3. Thus, for a beam with bending rigidity EI and torsional rigidity GJ, the strain energy becomes

$$U = \frac{1}{2} \int_0^\ell \left[EI \left(\frac{\partial^2 w}{\partial y^2} \right)^2 + GJ \left(\frac{\partial \theta}{\partial y} \right)^2 \right] dy. \tag{4.103}$$

To obtain the kinetic energy, we first consider the airfoil section shown in Fig. 3.12. Denoting the mass per unit volume of the material by ρ and noting that the velocity of a typical point within the cross-sectional plane is

$$\mathbf{v} = z \frac{\partial \theta}{\partial t} \hat{\mathbf{i}} + \left(\frac{\partial w}{\partial t} - x \frac{\partial \theta}{\partial t} \right) \hat{\mathbf{k}}, \tag{4.104}$$

where $\hat{\mathbf{i}}$ and $\hat{\mathbf{k}}$ are unit vectors in the x and z directions, respectively, one can write the kinetic energy as

$$K = \frac{1}{2} \int_0^\ell \iint_A \rho \left[\left(\frac{\partial w}{\partial t} - x \frac{\partial \theta}{\partial t} \right)^2 + z^2 \left(\frac{\partial \theta}{\partial t} \right)^2 \right] dx\, dz\, dy. \tag{4.105}$$

Straightforward evaluation of the cross-sectional integrals yields

$$K = \frac{1}{2} \int_0^\ell \left[m \left(\frac{\partial w}{\partial t} \right)^2 + 2md \frac{\partial w}{\partial t} \frac{\partial \theta}{\partial t} + mb^2 r^2 \left(\frac{\partial \theta}{\partial t} \right)^2 \right] dy, \qquad (4.106)$$

where m is the mass per unit length, d is the offset of the mass centroid from the elastic axis (positive when the mass centroid is toward the leading edge), and br is the cross-sectional mass radius of gyration about the elastic axis.

Finally, one needs the virtual work of the aerodynamic forces, which can be written as

$$\overline{\delta W} = \int_0^\ell [L'\delta w + (M'_{\rm ac} + eL')\delta\theta] \, dy, \qquad (4.107)$$

where, as before, L' and $M'_{\rm ac}$ are the distributed lift and pitching moment per unit length of the wing.

Owing to longstanding conventions in the literature of unsteady aerodynamics, this notation is not compatible with what we have used so far in this chapter. Thus, we rewrite these three expressions (i.e., strain energy, kinetic energy, and virtual work) in terms of the notation of this chapter. In particular, one can show that the following replacements can be made for the notation used in Fig. 3.12:

$$d \to -bx_\theta,$$

$$e \to \left(\frac{1}{2} + a \right) b,$$

$$M'_{\rm ac} \to M'_{\frac{1}{4}}.$$

Thus, the strain energy is unchanged from above. The kinetic energy becomes

$$K = \frac{1}{2} \int_0^\ell \left[m \left(\frac{\partial w}{\partial t} \right)^2 - 2mbx_\theta \frac{\partial w}{\partial t} \frac{\partial \theta}{\partial t} + mb^2 r^2 \left(\frac{\partial \theta}{\partial t} \right)^2 \right] dy, \qquad (4.108)$$

and the virtual work becomes

$$\overline{\delta W} = \int_0^\ell \left\{ L'\delta w + \left[M'_{\frac{1}{4}} + \left(\frac{1}{2} + a \right) bL' \right] \delta\theta \right\} dy. \qquad (4.109)$$

A reasonable choice for the assumed modes is the set of uncoupled free-vibration modes of the wing for bending and torsion, such that

$$w(y, t) = \sum_{i=1}^{N_w} \eta_i(t)\Psi_i(y),$$

$$\theta(y, t) = \sum_{i=1}^{N_\theta} \phi_i(t)\Theta_i(y), \qquad (4.110)$$

where N_w and N_θ are the numbers of modes used to represent bending and torsion, respectively; η_i and ϕ_i are the generalized coordinates associated with bending and torsion, respectively; and Ψ_i and Θ_i are the bending and torsion mode shapes, respectively. Here Θ_i is given by

$$\Theta_i = \sqrt{2} \sin(\gamma_i y), \qquad (4.111)$$

where

$$\gamma_i = \frac{\pi \left(i - \frac{1}{2}\right)}{\ell}, \tag{4.112}$$

and, according to Eq. (2.251), Ψ_i is given as

$$\Psi_i = \cosh(\alpha_i y) - \cos(\alpha_i y) - \beta_i[\sinh(\alpha_i y) - \sin(\alpha_i y)], \tag{4.113}$$

with α_i and β_i as given in Table 2.1.

The next step in the application of the Ritz method is to spatially discretize the strain energy, kinetic energy, and virtual work expressions. Because of orthogonality of the assumed modes, the strain energy simplifies to

$$U = \frac{1}{2}\left[\frac{EI}{\ell^3}\sum_{i=1}^{N_w}(\alpha_i\ell)^4\eta_i^2 + \frac{GJ}{\ell}\sum_{i=1}^{N_\theta}(\gamma_i\ell)^2\phi_i^2\right]. \tag{4.114}$$

The kinetic energy is also considerably simplified because of the orthogonality of the assumed modes and can be written as

$$K = \frac{m\ell}{2}\left(\sum_{i=1}^{N_w}\dot{\eta}_i^2 + b^2r^2\sum_{i=1}^{N_\theta}\dot{\phi}_i^2 - 2bx_\theta\sum_{i=1}^{N_\theta}\sum_{j=1}^{N_w}A_{ij}\dot{\phi}_i\dot{\eta}_j\right), \tag{4.115}$$

where

$$A_{ij} = \frac{1}{\ell}\int_0^\ell \Theta_i\Psi_j\,dy \qquad (i = 1, 2, \dots, N_\theta; \qquad j = 1, 2, \dots, N_w). \tag{4.116}$$

Note the inertial coupling between bending and torsion modes reflected by the term involving A_{ij}.

The virtual work expression can be used to identify the generalized forces

$$\overline{\delta W} = \sum_{i=1}^{N_w}\Xi_{w_i}\delta\eta_i + \sum_{i=1}^{N_\theta}\Xi_{\theta_i}\delta\phi_i. \tag{4.117}$$

Thus,

$$\Xi_{w_i} = \int_0^\ell \Psi_i L'\,dy,$$
$$\Xi_{\theta_i} = \int_0^\ell \Theta_i\left[M'_{\frac{1}{4}} + \left(\frac{1}{2}+a\right)bL'\right]dy, \tag{4.118}$$

where expressions for L' and $M'_{\frac{1}{4}}$ can be found by taking expressions for L and $M_{\frac{1}{4}}$ in Eqs. (4.78) or (4.95) and replacing h with $-w$ and dots with partial derivatives with respect to time. This we carry out for illustrative purposes using Theodorsen theory, for which

$$L' = 2\pi\rho_\infty UbC(k)\left[U\theta - \frac{\partial w}{\partial t} + b\left(\frac{1}{2}-a\right)\frac{\partial\theta}{\partial t}\right]$$
$$+ \pi\rho_\infty b^2\left(U\frac{\partial\theta}{\partial t} - \frac{\partial^2 w}{\partial t^2} - ba\frac{\partial^2\theta}{\partial t^2}\right), \tag{4.119}$$

$$M'_{\frac{1}{4}} = -\pi\rho_\infty b^3\left[U\frac{\partial\theta}{\partial t} - \frac{1}{2}\frac{\partial^2 w}{\partial t^2} + b\left(\frac{1}{8}-\frac{a}{2}\right)\frac{\partial^2\theta}{\partial t^2}\right].$$

Substituting Eqs. (4.110) into Eqs. (4.119), one obtains expressions for the generalized forces that can be easily put into matrix form:

$$\begin{Bmatrix} \Xi_w \\ \Xi_\theta \end{Bmatrix} = -\pi\rho_\infty b^2 \ell \begin{bmatrix} [\Delta] & ba[A]^T \\ ba[A] & b^2\left(a^2 + \frac{1}{8}\right)[\Delta] \end{bmatrix} \begin{Bmatrix} \ddot{\eta} \\ \ddot{\phi} \end{Bmatrix}$$

$$- \pi\rho_\infty bU\ell \begin{bmatrix} 2C(k)[\Delta] & -b\left[1 + 2\left(\frac{1}{2} - a\right)C(k)\right][A]^T \\ 2b\left(\frac{1}{2} + a\right)C(k)[A] & b^2\left(\frac{1}{2} - a\right)\left[1 - 2\left(\frac{1}{2} + a\right)C(k)\right][\Delta] \end{bmatrix} \begin{Bmatrix} \dot{\eta} \\ \dot{\phi} \end{Bmatrix}$$

$$- \pi\rho_\infty bU^2\ell \begin{bmatrix} [0] & -2C(k)[A]^T \\ [0] & -b(1 + 2a)C(k)[\Delta] \end{bmatrix} \begin{Bmatrix} \eta \\ \phi \end{Bmatrix},$$

$$(4.120)$$

where $[\Delta]$ denotes an identity matrix and $[0]$ denotes a matrix of zeros. Because of limitations inherent in the derivation of Theodorsen's theory, this expression for the generalized forces is valid only for simple harmonic motion.

All that now remains in the application of the Ritz method is to invoke Lagrange's equations to get the generalized equations of motion, which can be written in matrix form as

$$m\ell \begin{bmatrix} [\Delta] & -bx_\theta[A]^T \\ -bx_\theta[A] & b^2r^2[\Delta] \end{bmatrix} \begin{Bmatrix} \ddot{\eta} \\ \ddot{\phi} \end{Bmatrix} + \begin{bmatrix} \frac{EI}{\ell^3}[B] & [0] \\ [0] & \frac{GJ}{\ell}[T] \end{bmatrix} \begin{Bmatrix} \eta \\ \phi \end{Bmatrix} = \begin{Bmatrix} \Xi_w \\ \Xi_\theta \end{Bmatrix},$$

$$(4.121)$$

where elements of the diagonal matrices $[B]$ and $[T]$ are given by

$$B_{ii} = (\alpha_i\ell)^4,$$
$$T_{ii} = (\gamma_i\ell)^2. \tag{4.122}$$

The appearance of diagonal matrices $[B]$ and $[T]$ in the stiffness matrix and the appearances of Δ in the mass matrix and generalized forces are due to our choice of basis functions that are orthogonal. Such a choice is not necessary, but it does simplify the discretized equations.

Following the methodology of classical flutter analysis in Section 4.3, one sets

$$\eta(t) = \bar{\eta}\exp(i\omega t),$$
$$\phi(t) = \bar{\phi}\exp(i\omega t), \tag{4.123}$$

where ω is the frequency of the simple harmonic motion. This leads to a flutter determinant that can be solved by following steps similar to those outlined in Section 4.3, the only difference being that there are now more degrees of freedom if either N_w or N_θ exceeds unity.

Let us consider the case in which $N_w = N_\theta = 1$. If we introduce dimensionless constants similar to those in Section 4.3, the equations of motion can be put in the form of Eqs. (4.58), that is,

$$\left\{\mu\left[1 - \left(\frac{\omega_w}{\omega}\right)^2\right] + \ell_w\right\}\frac{\bar{\eta}_1}{b} + (-\mu x_\theta + \ell_\theta)A_{11}\bar{\phi}_1 = 0,$$

$$(-\mu x_\theta + m_w)A_{11}\frac{\bar{\eta}_1}{b} + \left\{\mu r^2\left[1 - \left(\frac{\omega_\theta}{\omega}\right)^2\right] + m_\theta\right\}\bar{\phi}_1 = 0. \tag{4.124}$$

Here ℓ_w, ℓ_θ, m_w, and m_θ are defined in a manner similar to the quantities on the right-hand side of Eqs. (4.35) with the loads from Theodorsen theory,

$$\ell_w = 1 - \frac{2i\,C(k)}{k},$$

$$\ell_\theta = a + \frac{i}{k}\left[1 + 2\left(\frac{1}{2} - a\right)C(k)\right] + \frac{2C(k)}{k^2},$$

$$m_w = a - \frac{2i\left(\frac{1}{2} + a\right)C(k)}{k},$$

$$m_\theta = a^2 + \frac{1}{8} - \frac{\left(\frac{1}{2} - a\right)\left[1 - 2\left(\frac{1}{2} + a\right)C(k)\right]i}{k} + \frac{2\left(\frac{1}{2} + a\right)C(k)}{k^2},$$

(4.125)

and the fundamental bending and torsion frequencies are

$$\omega_w = (\alpha_1\ell)^2\sqrt{\frac{EI}{m\ell^4}},$$

$$\omega_\theta = \frac{\pi}{2}\sqrt{\frac{GJ}{mb^2r^2\ell^2}}.$$

(4.126)

Finally, the constant $A_{11} = 0.958641$. It is clear that these equations are in the same form as the ones solved earlier for the typical section and that the influence of wing flexibility for this simplest two-mode case only enters in a minor way, namely, to adjust the coupling terms by a factor of less than 5%.

The main purpose of this example is to demonstrate how the tools already presented can be used to conduct a flutter analysis of a flexible wing. Addition of higher modes can certainly affect the results, as can such things as spanwise variations in the mass and stiffness properties and concentrated masses and inertias along the wing. Incorporation of these additional features into the analysis would serve to make the analysis more suitable for realistic flutter calculations. However, to fully capture the realism afforded by these and other important considerations, such as aircraft with delta-wing configurations or very low aspect ratio wings, a full finite element analysis would be necessary. Even in such cases it is typical that flutter analyses based on assumed modes give the analyst a reasonably good idea of the mechanisms of instability. Moreover, the full finite element method can be used to obtain a realistic set of assumed modes that could, in turn, be used in a Ritz analysis similar to that above.

4.7 Flutter Boundary Characteristics

The preceding sections have described procedures for the determination of the flutter boundary in terms of altitude, speed, and Mach number. For a standard atmosphere any two of these conditions is sufficient to describe the flight condition. The final flutter boundary is usually presented in terms of a dimensionless flutter speed as $U_F/(b\omega_\theta)$. This parameter is similar to the reciprocal of reduced frequency and is called the reduced velocity. A useful presentation of this reduced flutter speed as a function of the mass ratio, $\mu = m/(\pi\rho_\infty b^2)$, is illustrated in Fig. 4.13. It is immediately apparent that the flutter speed increases in a nearly linear fashion with increasing mass ratio. This result can be interpreted in either of two ways. For a given configuration, variations in μ would correspond to changes in atmospheric density and thus altitude. In such a case the mass ratio increases with increasing altitude.

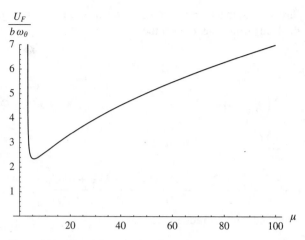

Figure 4.13 Plot of dimensionless flutter speed versus mass ratio for the case $\sigma = 1/\sqrt{10}, r = 1/2$, $x_\theta = 0$, and $a = -3/10$.

This implies that any flight vehicle is more susceptible to aeroelastic flutter at low altitudes than it is at higher ones.

A second interpretation of the mass ratio is related to its numerical value for any fixed altitude. The value of μ will depend on the type of flight vehicle as reflected by its mass per unit span, m. Table 4.2 gives some vehicle configurations and typical mass ratio values for atmospheric densities between sea level and 10,000 ft.

The flutter boundary is very sensitive to the dimensionless parameters. In Fig. 4.14, for example, one sees a dramatic change in the flutter speed versus the frequency ratio $\sigma = \omega_h/\omega_\theta$. The significant drop in the flutter speed for $x_\theta = 0.2$ around $\sigma = 1.4$ is of utmost practical importance. There are certain frequency ratios at which the flutter speed becomes very small, depending on the values of the other parameters. This dip will be observed in the plot of flutter speed versus frequency ratio for the wings of most high-performance aircraft, which have relatively large mass ratios and positive static unbalances. The chordwise offsets also have a strong influence on the flutter speed, as seen in Fig. 4.15. Indeed, a small change in the mass center location can lead to a large increase in the flutter speed. The mass center location, e, cannot be changed without simultaneously changing the dimensionless radius of gyration, r, but the relative change in the flutter speed for a small percentage change in the former is more than for a similar percentage change in the latter. These facts have led to a concept of mass balancing wings to alleviate flutter, similar to the way control surfaces are mass balanced. If the center of mass is moved forward of the reference point, the flutter speed is generally relatively high. Unfortunately, however, this

Table 4.2. *Variation of Mass Ratio for Typical Vehicle Types*

Vehicle Type	$\mu = \dfrac{m}{\pi \rho_\infty b^2}$
Gliders and Ultralights	5–15
General Aviation	10–20
Commercial Transports	15–30
Attack Aircraft	25–55
Helicopter Blades	65–110

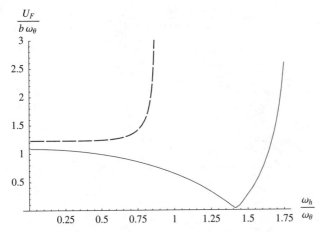

Figure 4.14   Plot of dimensionless flutter speed versus frequency ratio for the case $\mu = 3$, $r = 1/2$, and $a = -1/5$ where the solid line is for $x_\theta = 0.2$ and the dashed line is for $x_\theta = 0.1$.

is not easily accomplished, but fortunately a large change is not usually needed to ensure safety. It is noted that care must be exercised in examining changes in other parameters caused by such changes in the mass distribution. For example, the torsional frequency may be altered significantly in the process of changing the radius of gyration. Finally, we note that the flutter frequency for bending–torsion flutter is somewhere in between ω_h and ω_θ, where normally $\sigma < 1$. However, situations do arise in which the flutter frequency may exceed ω_θ.

It is important to note there are some combinations of the chordwise offset parameters e and a for which the present simplified theories indicate that flutter is not possible. Bisplinghoff, Ashley, and Halfman (1955) classify the effects of the chordwise offsets e and a in terms of small and large σ. For small σ, they note that flutter can only happen when the mass center is behind the quarter-chord (i.e., when $e > -1/2$) and thus it cannot happen when $e \leq -1/2$. For large σ, flutter can only happen when the elastic axis is in front of the quarter-chord (i.e., when $a < -1/2$) and thus it cannot happen when $a \geq -1/2$. Moreover, for the typical section model in combination with the aerodynamic models presented herein,

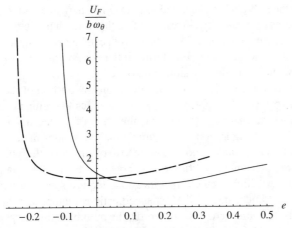

Figure 4.15   Plot of dimensionless flutter speed versus e for the case $\mu = 10$, $\sigma = 1/\sqrt{2}$, and $r = 1/2$; the solid line is for $a = 0$ and the dashed line is for $a = 0.2$.

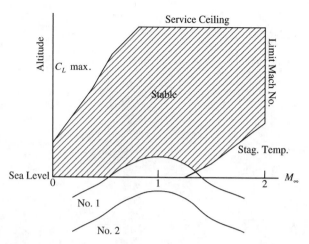

Figure 4.16   Flight envelope for typical Mach 2 fighter.

flutter does not appear to happen for any combination of σ and r when the mass centroid, elastic axis, and aerodynamic center all coincide (i.e., when $e = a = -1/2$). Even if this prediction of the analysis is correct, practically speaking, it is very difficult to achieve coincidence of these points in wing design. Remember, however, that all these statements are made with respect to simplified models. One needs to analyze real wings in a design setting using powerful tools, such as NASTRAN or ASTROS. Indeed, bending–torsion flutter is a very complicated phenomenon, and it seems to defy all our attempts at generalization. Additional discussion of these phenomena, along with a large body of solution plots, can be found in Bisplinghoff, Ashley, and Halfman (1955).

The final flutter boundary can be presented in numerous ways for any given flight vehicle. The manner in which it is illustrated depends on the engineering purpose it is intended to serve. One possible presentation of the flutter boundary is to superpose it on the vehicle's flight envelope. A typical flight envelope for a Mach 2 attack aircraft is illustrated in Fig. 4.16 with two flutter boundaries indicated by the curves marked "No. 1" and "No. 2." The shaded region above the flutter boundaries, being at higher altitudes, corresponds to stable flight conditions; below the boundaries flutter will be experienced. Flutter boundary no. 1 indicates that for a portion of the intended flight envelope, the vehicle will experience flutter. It should be noted that these conditions of instability correspond to a flight Mach number near unity (transonic flow) and high dynamic pressure. This observation can be generalized by saying that a flight vehicle is more susceptible to aeroelastic flutter for conditions of (1) lower altitude, (2) transonic flow, and (3) higher dynamic pressure.

If it is determined that the vehicle will experience flutter in any portion of its intended flight envelope, it is necessary to make appropriate design changes to eliminate the instability for such conditions. These changes may involve alteration of the inertial, elastic, or aerodynamic properties of the configuration. Many times small variations in all three provide the best compromise. Flutter boundary no. 2 is indicative of a flutter-safe vehicle. Note that at the minimum altitude transonic condition there appears to be a safety margin with respect to flutter instability. All flight vehicle specifications require such a safety factor, which is generally called the "flutter margin." Most specifications require that the margin be 15% over the limit equivalent airspeed. In other words, the minimum flutter speed at sea level should not be less that 1.15 times the airspeed for the maximum expected dynamic pressure as evaluated at sea level.

4.8   Epilogue

In this chapter we have considered the general problem of lifting-surface flutter. Several types of flutter analysis have been presented, including the p method, classical flutter analysis, the k method, and the p–k method. The application of classical flutter analysis to discrete one- and two-degree-of-freedom wind tunnel models has been presented. The student has been exposed to Theodorsen's unsteady thin-airfoil theory along with the more modern finite-state thin-airfoil theory of Peters et al. Application of the assumed modes method to construct a flutter analysis of a flexible wing has been demonstrated as well. Finally, some of the important parameters of the flutter problem have been discussed, along with current design practice. With a good understanding of the material presented herein, the student should be sufficiently equipped to apply these fundamentals to the design of flight vehicles.

Moreover, with appropriate graduate-level studies well beyond the scope of material presented herein, the student will be able conduct research in the exciting field of aeroelasticity. Current research topics are quite diverse. With the increase in the sophistication of controls technology, it has become more and more common to attack flutter problems by active control of flaps or other flight control surfaces. These so-called flutter suppression systems provide alternatives to costly design changes. One type of system for which flutter suppression systems are an excellent choice is a military aircraft that must carry weapons as stores. These aircraft must be free of flutter within their flight envelope for several different configurations. Sometimes avoidance of flutter by design changes is simply beyond the capability of the designer for such complex systems. There is also research to determine in flight when a flutter boundary is being approached. This could be of great value for situations in which damage had altered the properties of the aircraft structure, perhaps unknown to the pilot, thus shifting the flutter (or divergence) boundary and making the aircraft unsafe to operate within its original flight envelope. Other current problems of interest to aeroelasticians include improved analysis methodology for prediction of flutter, gust response, and limit-cycle oscillations, design of control systems to improve gust response and limit-cycle oscillations; and incorporation of aeroelastic analyses at an earlier stage of aircraft design.

Problems

1. Compute the flutter speed for the incompressible, one-degree-of-freedom flutter problem with

$$m_\theta = \frac{i-2}{k} - 10i,$$

$$I_P = 50\pi\rho_\infty b^4, \qquad \omega_\theta = 10 \text{ Hz}, \qquad b = 0.5 \text{ ft}.$$

Answer: $U_F = 405.6$ ft/s

2. According to Theodorsen's theory, the circulatory lift is proportional to a quantity that, for simple harmonic motion, can be shown to be equal to the effective angle of attack given by

$$\alpha = C(k)\left[\theta + \frac{\dot{h}}{U} + \frac{b}{U}\left(\frac{1}{2} - a\right)\dot{\theta}\right].$$

For $a = -1/2$ and simple harmonic motion such that $\bar{\theta} = 1$ and $\bar{h} = bz[\cos(\phi) + i\sin(\phi)]$, plot α as a function of time for five periods for the following four cases:

(a) $z = 0.1$; $\phi = 0°$; $k = 0.01$
(b) $z = 0.1$; $\phi = 0°$; $k = 1.0$
(c) $z = 0.1$; $\phi = 90°$; $k = 1.0$
(d) $z = 0.5$; $\phi = 90°$; $k = 1.0$

Comment on the behavior of α for increasing k, changing the phase angle from $0°$ to $90°$, and increasing the plunge magnitude. You may approximate Theodorsen's function as

$$C(k) = \frac{0.01365 + 0.2808ik - \frac{k^2}{2}}{0.01365 + 0.3455ik - k^2}.$$

3. Show that the coefficients used in a classical flutter analysis, if based on Theodorsen's theory, are

$$\ell_h = 1 - \frac{2iC(k)}{k},$$

$$\ell_\theta = -a - \frac{i}{k} - \frac{2C(k)}{k^2} - \frac{2i\left(\frac{1}{2} - a\right)C(k)}{k},$$

$$m_h = -a + \frac{2i\left(\frac{1}{2} + a\right)C(k)}{k},$$

$$m_\theta = \frac{1}{8} + a^2 - \frac{i\left(\frac{1}{2} - a\right)}{k} + \frac{2\left(\frac{1}{2} + a\right)C(k)}{k^2} + \frac{2i\left(\frac{1}{4} - a^2\right)C(k)}{k}.$$

4. Consider an incompressible, two-degree-of-freedom flutter problem in which $a = -1/5$, $e = -1/10$, $\mu = 20$, $r^2 = 6/25$, and $\sigma = 2/5$. Compute the flutter speed and the flutter frequency using the classical flutter approach. For the aerodynamic coefficients use those of Theodorsen's theory with $C(k)$ approximated as in Problem 2.

Answers: $U_F = 2.170\, b\omega_\theta$ and $\omega_F = 0.6443\, \omega_\theta$

5. Consider an incompressible, two-degree-of-freedom flutter problem in which $a = -1/5$, $\mu = 3$, and $r = 1/2$. Compute the flutter speed and flutter frequency for two cases $x_\theta = e - a = 1/5$ and $x_\theta = e - a = 1/10$, and let $\sigma = 0.2, 0.4, 0.6, 0.8$, and 1.0. Use the classical flutter approach, and for the aerodynamic coefficients use those of Theodorsen's theory with $C(k)$ approximated as in Problem 2. Compare with the results in Fig. 4.14.

6. Set up the complete set of equations for flutter analysis by the p method using the unsteady aerodynamic theory of Peters et al. (1995), nondimensionalizing Eqs. (4.97) and redefining λ_i as $b\omega_\theta\lambda_i$.

7. Write a computer program using MATLAB or Mathematica to set up the solution of the equations derived in Problem 6.

8. Using the computer program written in Problem 7, solve for the dimensionless flutter speed and flutter frequency for an incompressible, two-degree-of-freedom flutter problem in which $a = -1/3$, $e = -1/10$, $\mu = 50$, $r = 2/5$, and $\sigma = 2/5$.

Answers: $U_F = 2.807 b\omega_\theta$ and $\omega_F = 0.5952\omega_\theta$

9. Write a computer program using MATLAB or Mathematica to set up the solution of a two-degree-of-freedom flutter problem using the k method.

10. Use the computer program written in Problem 9 to solve a flutter problem in which $a = -1/5$, $e = -1/10$, $\mu = 20$, $r^2 = 6/25$, and $\sigma = 2/5$. Plot the values of $\omega_{1,2}/\omega_\theta$ and g versus $U/(b\omega_\theta)$ and compare your results with the quantities plotted in Figs. 4.11 and 4.12. Noting how the quantities plotted in these two sets of figures are different, comment on the similarities and differences you observe in these plots and why those differences are there. Finally, explain why your predicted flutter speed is the same as that determined with the classical method.

Answer: See Figs. 4.17 and 4.18.

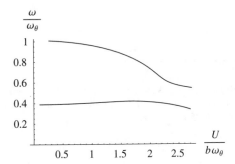

Figure 4.17 Plot of $\omega_{1,2}/\omega_\theta$ versus $U/(b\omega_\theta)$ using the k method and Theodorsen aerodynamics with $a = -1/5$, $e = -1/10$, $\mu = 20$, $r^2 = 6/25$, and $\sigma = 2/5$.

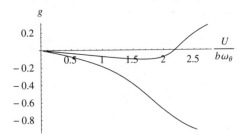

Figure 4.18 Plot of g versus $U/(b\omega_\theta)$ using the k method and Theodorsen aerodynamics with $a = -1/5$, $e = -1/10$, $\mu = 20$, $r^2 = 6/25$, and $\sigma = 2/5$.

11. Show that the flutter determinant for the p–k method applied to the typical section using Theodorsen aerodynamics can be expressed as

$$
\begin{bmatrix}
p^2 + \dfrac{\sigma^2}{V^2} - \dfrac{k^2}{\mu} + \dfrac{2ikC(k)}{\mu} & \dfrac{p^2\mu x_\theta + k(i+ak) + [2+ik(1-2a)]C(k)}{\mu} \\[2ex]
\dfrac{p^2\mu x_\theta + ak^2 - ik(1+2a)C(k)}{\mu} & r^2\left(p^2 + \dfrac{1}{V^2}\right) + \dfrac{4i(1+2a)[2i-k(1-2a)]C(k) - k[k-4i+8a(i+ak)]}{8\mu}
\end{bmatrix}
$$

12. Write a computer program using MATLAB or Mathematica to set up the solution of a two-degree-of-freedom flutter problem using the p–k method and Theodorsen aerodynamics.

13. Use the computer program written in Problem 12 to solve a flutter problem in which $a = -1/5$, $e = -1/10$, $\mu = 20$, $r^2 = 6/25$, and $\sigma = 2/5$. Plot the values of the estimates of $\Omega_{1,2}/\omega_\theta$ and $\Gamma_{1,2}/\omega_\theta$ versus $U/(b\omega_\theta)$ and compare your results with the quantities plotted in Figs. 4.11 and 4.12. Explain why the estimated damping from the p–k method sometimes differs from that of the p method.

Answer: See Figs. 4.19 and 4.20.

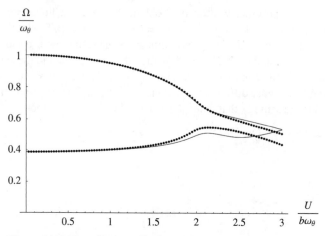

Figure 4.19 Plot of estimated value of $\Omega_{1,2}/\omega_\theta$ versus $U/(b\omega_\theta)$ using the $p-k$ method with Theodorsen aerodynamics (symbols) and the p method with the aerodynamics of Peters et al. (lines) for $a = -1/5$, $e = -1/10$, $\mu = 20$, $r^2 = 6/25$, and $\sigma = 2/5$.

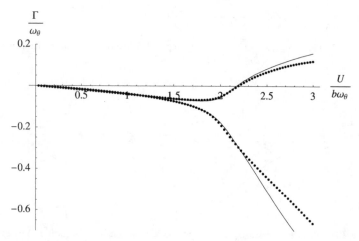

Figure 4.20 Plot of estimated value of $\Gamma_{1,2}/\omega_\theta$ versus $U/(b\omega_\theta)$ using the $p-k$ method with Theodorsen aerodynamics (symbols) and the p method with the aerodynamics of Peters et al. (lines) for $a = -1/5$, $e = -1/10$, $\mu = 20$, $r^2 = 6/25$, and $\sigma = 2/5$.

14. Write a computer program using MATLAB or Mathematica to set up the solution of a two-degree-of-freedom flutter problem using the $p-k$ method and the aerodynamics of Peters et al.

15. Using the computer programs of Problems 12 and 14, show that the $p-k$ method yields the same results regardless of whether one uses Theodorsen's theory or the aerodynamic theory of Peters et al., assuming a sufficiently large number of inflow states is used in the latter. You may do this for the case $a = -1/5$, $e = -1/10$, $\mu = 20$, $r^2 = 6/25$, and $\sigma = 2/5$. What does this imply about the two theories?

Lagrange's Equation

A.1   Introduction

When we wish to use Newton's Laws to write the equations of motion of a particle or a system of particles we must be careful to include all the forces of the system. The Lagrangean form of the equations of motion that we shall proceed to derive has the advantage that we can ignore all forces that do no work (e.g., forces at frictionless pins, forces at a point of rolling contact, forces at frictionless guides, and forces in inextensible connections). In the case of conservative systems (systems for which the total energy remains constant) the Lagrangean method gives us an automatic procedure for obtaining the equations of motion provided only that we can write the kinetic and potential energies of the system.

A.2   Degrees of Freedom

Before proceeding to develop the Lagrange equations we must characterize our dynamical systems on some systematic way. The most important property of this sort for our present purpose is the number of independent coordinates that we must know to completely specify the position or configuration of our system. We say that a system has n degrees of freedom if exactly n coordinates serve to completely define its configuration.

Example 1

A free particle in space has three degrees of freedom since we must know three coordinates, x, y, z, for example, to locate it.

Example 2

A wheel that rolls without slipping on a straight track has one degree of freedom since either the distance from some base point or the total angle of rotation will enable us to locate it completely.

A.3   Generalized Coordinates

We usually think of coordinates as lengths or angles. However, any set of parameters that enable us to uniquely specify the configuration of the system can serve as coordinates. When we generalize the meaning of the term in this manner, we call these new quantities generalized coordinates.

Example 3

Consider a bar rotating in a plane about a point O. The angle of rotation with respect to some base line suggests itself as an obvious coordinate for specifying the position of the bar. However, the area swept over by the bar would do equally well and could therefore be used as a generalized coordinate.

If a system has n degrees of freedom, then n generalized coordinates are necessary and sufficient to determine its configuration.

155

A.4   Lagrange's Equations

In deriving these equations we shall consider systems having two degrees of free-dom and hence are completely defined by two generalized coordinates q_1 and q_2. However, the results are easily extended to systems having any number of degrees of freedom.

Suppose our system is made up of n particles. For each particle we can write by Newton's Second Law

$$
\begin{aligned}
M_i \ddot{x}_i &= X_i, \\
M_i \ddot{y}_i &= Y_i, \\
M_i \ddot{z}_i &= Z_i,
\end{aligned}
\tag{A.1}
$$

where x_i, y_i, z_i are the rectangular Cartesian coordinates of the ith particle, M_i is its mass, and X_i, Y_i, Z_i are the resultants of all forces acting on it in the x, y, and z directions, respectively.

If we multiply both sides of Eqs. (A.1) by δx_i, δy_i, and δz_i, respectively, and add the equations we have

$$
M_i \left(\ddot{x}_i \delta x_i + \ddot{y}_i \delta y_i + \ddot{z}_i \delta z_i \right) = X_i \delta x_i + Y_i \delta y_i + Z_i \delta z_i.
\tag{A.2}
$$

The right-hand side of this equation represents the work done by all the forces acting on the ith particle during the virtual displacements δx_i, δy_i, and δz_i. Hence forces that do no work do not contribute to the right-hand side of Eq. (A.2) and may be omitted from the equation. To obtain the corresponding equation for the whole system, we sum both sides of Eq. (A.2) for all particles. Thus

$$
\sum_{i=1}^{n} M_i \left(\ddot{x}_i \delta x_i + \ddot{y}_i \delta y_i + \ddot{z}_i \delta z_i \right) = \sum_{i=1}^{n} \left(X_i \delta x_i + Y_i \delta y_i + Z_i \delta z_i \right).
\tag{A.3}
$$

Now because our system is completely located in space if we know the two generalized coordinates q_1 and q_2, we must be able to write x_i, y_i, and z_i as well as their increments δx_i, δy_i, and δz_i as functions of q_1 and q_2. Hence,

$$
\begin{aligned}
x_i &= x_i(q_1, q_2), \\
y_i &= y_i(q_1, q_2), \\
z_i &= z_i(q_1, q_2).
\end{aligned}
\tag{A.4}
$$

Differentiating Eq. (A.4) with respect to time gives

$$
\begin{aligned}
\dot{x}_i &= \frac{\partial x_i}{\partial q_1} \dot{q}_1 + \frac{\partial x_i}{\partial q_2} \dot{q}_2, \\
\dot{y}_i &= \frac{\partial y_i}{\partial q_1} \dot{q}_1 + \frac{\partial y_i}{\partial q_2} \dot{q}_2, \\
\dot{z}_i &= \frac{\partial z_i}{\partial q_1} \dot{q}_1 + \frac{\partial z_i}{\partial q_2} \dot{q}_2.
\end{aligned}
\tag{A.5}
$$

Similarly,

$$\delta x_i = \frac{\partial x_i}{\partial q_1}\delta q_1 + \frac{\partial x_i}{\partial q_2}\delta q_2,$$

$$\delta y_i = \frac{\partial y_i}{\partial q_1}\delta q_1 + \frac{\partial y_i}{\partial q_2}\delta q_2, \tag{A.6}$$

$$\delta z_i = \frac{\partial z_i}{\partial q_1}\delta q_1 + \frac{\partial z_i}{\partial q_2}\delta q_2.$$

If we substitute these into Eq. (A.3) and rearrange the terms we obtain

$$\sum_{i=1}^{n}\left[M_i\left(\ddot{x}_i\frac{\partial x_i}{\partial q_1} + \ddot{y}_i\frac{\partial y_i}{\partial q_1} + \ddot{z}_i\frac{\partial z_i}{\partial q_1}\right)\delta q_1 \right.$$

$$\left. + M_i\left(\ddot{x}_i\frac{\partial x_i}{\partial q_2} + \ddot{y}_i\frac{\partial y_i}{\partial q_2} + \ddot{z}_i\frac{\partial z_i}{\partial q_2}\right)\delta q_2 \right]$$

$$= \sum_{i=1}^{n}\left[\left(X_i\frac{\partial x_i}{\partial q_1} + Y_i\frac{\partial y_i}{\partial q_1} + Z_i\frac{\partial z_i}{\partial q_1}\right)\delta q_1 \right.$$

$$\left. + \left(X_i\frac{\partial x_i}{\partial q_2} + Y_i\frac{\partial y_i}{\partial q_2} + Z_i\frac{\partial z_i}{\partial q_2}\right)\delta q_2 \right]. \tag{A.7}$$

From Eq. (A.5) we conclude that since x_i, y_i, and z_i are functions of q_1 and q_2 but not of $\dot{q}_1$ and $\dot{q}_2$

$$\frac{\partial \dot{x}_i}{\partial \dot{q}_1} = \frac{\partial x_i}{\partial q_1}, \qquad \frac{\partial \dot{x}_i}{\partial \dot{q}_2} = \frac{\partial x_i}{\partial q_2},$$

$$\frac{\partial \dot{y}_i}{\partial \dot{q}_1} = \frac{\partial y_i}{\partial q_1}, \qquad \frac{\partial \dot{y}_i}{\partial \dot{q}_2} = \frac{\partial y_i}{\partial q_2}, \tag{A.8}$$

$$\frac{\partial \dot{z}_i}{\partial \dot{q}_1} = \frac{\partial z_i}{\partial q_1}, \qquad \frac{\partial \dot{z}_i}{\partial \dot{q}_2} = \frac{\partial z_i}{\partial q_2}.$$

We substitute these relations into the left-hand side of Eq. (A.7) to obtain

$$\sum_{i=1}^{n}\left[M_i\left(\ddot{x}_i\frac{\partial \dot{x}_i}{\partial \dot{q}_1} + \ddot{y}_i\frac{\partial \dot{y}_i}{\partial \dot{q}_1} + \ddot{z}_i\frac{\partial \dot{z}_i}{\partial \dot{q}_1}\right)\delta q_1 \right.$$

$$\left. + M_i\left(\ddot{x}_i\frac{\partial \dot{x}_i}{\partial \dot{q}_2} + \ddot{y}_i\frac{\partial \dot{y}_i}{\partial \dot{q}_2} + \ddot{z}_i\frac{\partial \dot{z}_i}{\partial \dot{q}_2}\right)\delta q_2 \right]$$

$$= \sum_{i=1}^{n}\left[\left(X_i\frac{\partial x_i}{\partial q_1} + Y_i\frac{\partial y_i}{\partial \dot{q}_1} + Z_i\frac{\partial z_i}{\partial q_1}\right)\delta q_1 \right. \tag{A.9}$$

$$\left. + \left(X_i\frac{\partial x_i}{\partial q_2} + Y_i\frac{\partial y_i}{\partial q_2} + Z_i\frac{\partial z_i}{\partial q_2}\right)\delta q_2 \right].$$

Now let us shift our attack on the problem and consider the kinetic energy of the system. This is

$$K = \frac{1}{2}\sum_{i=1}^{n} M_i\left(\dot{x}_i^2 + \dot{y}_i^2 + \dot{z}_i^2\right). \tag{A.10}$$

Now calculate $\frac{\partial K}{\partial \dot{q}_1}$ and $\frac{\partial K}{\partial \dot{q}_2}$ to obtain

$$\frac{\partial K}{\partial \dot{q}_1} = \sum_{i=1}^{n} M_i \left(\dot{x}_i \frac{\partial \dot{x}_i}{\partial \dot{q}_1} + \dot{y}_i \frac{\partial \dot{y}_i}{\partial \dot{q}_1} + \dot{z}_i \frac{\partial \dot{z}_i}{\partial \dot{q}_1} \right),$$ (A.11)

$$\frac{\partial K}{\partial q_1} = \sum_{i=1}^{n} M_i \left(\dot{x}_i \frac{\partial \dot{x}_i}{\partial q_1} + \dot{y}_i \frac{\partial \dot{y}_i}{\partial q_1} + \dot{z}_i \frac{\partial \dot{z}_i}{\partial q_1} \right).$$ (A.12)

We next calculate the time derivative of $\frac{\partial x_i}{\partial q_1}$, for which the chain rule gives

$$\frac{d}{dt} \left(\frac{\partial x_i}{\partial q_1} \right) = \frac{\partial^2 x_i}{\partial q_1^2} \dot{q}_1 + \frac{\partial^2 x_i}{\partial q_1 \partial q_2} \dot{q}_2$$

$$= \frac{\partial}{\partial q_1} \left(\frac{\partial x_i}{\partial q_1} \dot{q}_1 + \frac{\partial x_i}{\partial q_2} \dot{q}_2 \right)$$

$$= \frac{\partial}{\partial q_1} (\dot{x}_i) = \frac{\partial \dot{x}_i}{\partial q_1}.$$ (A.13)

Since from Eq. (A.8) we have

$$\frac{\partial \dot{x}_i}{\partial \dot{q}_1} = \frac{\partial x_i}{\partial q_1},$$ (A.14)

we conclude from Eq. (A.13) that

$$\frac{d}{dt} \left(\frac{\partial x_i}{\partial q_1} \right) = \frac{\partial \dot{x}_i}{\partial q_1}.$$ (A.15)

The following relations can be proven in a similar manner:

$$\frac{d}{dt} \left(\frac{\partial y_i}{\partial q_1} \right) = \frac{\partial \dot{y}_i}{\partial q_1},$$

$$\frac{d}{dt} \left(\frac{\partial z_i}{\partial q_1} \right) = \frac{\partial \dot{z}_i}{\partial q_1}.$$ (A.16)

Now let us use Eqs. (A.11), (A.12), (A.15), and (A.16) to calculate the function

$$\frac{d}{dt} \left(\frac{\partial K}{\partial \dot{q}_1} \right) - \frac{\partial K}{\partial q_1},$$ (A.17)

for which the result is

$$\frac{d}{dt} \left(\frac{\partial K}{\partial \dot{q}_1} \right) - \frac{\partial K}{\partial q_1} = \sum_{i=1}^{n} M_i \left(\ddot{x}_i \frac{\partial \dot{x}_i}{\partial \dot{q}_1} + \ddot{y}_i \frac{\partial \dot{y}_i}{\partial \dot{q}_1} + \ddot{z}_i \frac{\partial \dot{z}_i}{\partial \dot{q}_1} \right)$$

$$+ \sum_{i=1}^{n} M_i \left[\dot{x}_i \frac{d}{dt} \left(\frac{\partial \dot{x}_i}{\partial \dot{q}_1} \right) + \dot{y}_i \frac{d}{dt} \left(\frac{\partial \dot{y}_i}{\partial \dot{q}_1} \right) + \dot{z}_i \frac{d}{dt} \left(\frac{\partial \dot{z}_i}{\partial \dot{q}_1} \right) \right]$$

$$- \sum_{i=1}^{n} M_i \left(\dot{x}_i \frac{\partial \dot{x}_i}{\partial q_1} + \dot{y}_i \frac{\partial \dot{y}_i}{\partial q_1} + \dot{z}_i \frac{\partial \dot{z}_i}{\partial q_1} \right).$$ (A.18)

From Eqs. (A.15) and (A.16) the second and third terms on the right-hand side of Eq. (A.18) are equal and thus cancel, leaving

$$\frac{d}{dt}\left(\frac{\partial K}{\partial \dot{q}_1}\right) - \frac{\partial K}{\partial q_1} = \sum_{i=1}^{n} M_i \left(\ddot{x}_i \frac{\partial \dot{x}_i}{\partial \dot{q}_1} + \ddot{y}_i \frac{\partial \dot{y}_i}{\partial \dot{q}_1} + \ddot{z}_i \frac{\partial \dot{z}_i}{\partial \dot{q}_1} \right). \tag{A.19}$$

A similar relation holds for partial derivatives of K with respect to q_2 and $\dot{q}_2$. Hence Eq. (A.9) can be written

$$\left[\frac{d}{dt}\left(\frac{\partial K}{\partial \dot{q}_1}\right) - \frac{\partial K}{\partial q_1}\right]\delta q_1 + \left[\frac{d}{dt}\left(\frac{\partial K}{\partial \dot{q}_2}\right) - \frac{\partial K}{\partial q_2}\right]\delta q_2$$

$$= \sum_{i=1}^{n} \left(X_i \frac{\partial x_i}{\partial q_1} + Y_i \frac{\partial y_i}{\partial q_1} + Z_i \frac{\partial z_i}{\partial q_1} \right)\delta q_1$$

$$+ \sum_{i=1}^{n} \left(X_i \frac{\partial x_i}{\partial q_2} + Y_i \frac{\partial y_i}{\partial q_2} + Z_i \frac{\partial z_i}{\partial q_2} \right)\delta q_2. \tag{A.20}$$

Since q_1 and q_2 are independent coordinates they can be varied arbitrarily. Hence, we can conclude that

$$\frac{d}{dt}\left(\frac{\partial K}{\partial \dot{q}_1}\right) - \frac{\partial K}{\partial q_1} = \sum_{i=1}^{n} \left(X_i \frac{\partial x_i}{\partial q_1} + Y_i \frac{\partial y_i}{\partial q_1} + Z_i \frac{\partial z_i}{\partial q_1} \right),$$

$$\frac{d}{dt}\left(\frac{\partial K}{\partial \dot{q}_2}\right) - \frac{\partial K}{\partial q_2} = \sum_{i=1}^{n} \left(X_i \frac{\partial x_i}{\partial q_2} + Y_i \frac{\partial y_i}{\partial q_2} + Z_i \frac{\partial z_i}{\partial q_2} \right). \tag{A.21}$$

The right-hand side of Eq. (A.20) is the work done by all the forces on the system when the coordinates of the ith particle undergo the small displacement δx_i, δy_i, and δz_i due to changes δq_1 and δq_2 in the generalized coordinates q_1 and q_2. The coefficients of δq_1 and δq_2 are known as the generalized forces Q_1 and Q_2, since they are the quantities by which the variations of the generalized coordinates must be multiplied to calculate the virtual work done by all the forces acting on the system. Hence,

$$Q_1 = \sum_{i=1}^{n} \left(X_i \frac{\partial x_i}{\partial q_1} + Y_i \frac{\partial y_i}{\partial q_1} + Z_i \frac{\partial z_i}{\partial q_1} \right),$$

$$Q_2 = \sum_{i=1}^{n} \left(X_i \frac{\partial x_i}{\partial q_2} + Y_i \frac{\partial y_i}{\partial q_2} + Z_i \frac{\partial z_i}{\partial q_2} \right), \tag{A.22}$$

and Eqs. (A.21) can be written

$$\frac{d}{dt}\left(\frac{\partial K}{\partial \dot{q}_1}\right) - \frac{\partial K}{\partial q_1} = Q_1,$$

$$\frac{d}{dt}\left(\frac{\partial K}{\partial \dot{q}_2}\right) - \frac{\partial K}{\partial q_2} = Q_2. \tag{A.23}$$

This is one form of Lagrange's equations of motion. They apply to any system that is completely described by two and only two generalized coordinates, whether or not the system is conservative. It can be shown by slightly more extended calculation that they apply to systems of any finite number of degrees of freedom.

A.5   Lagrange's Equations for Conservative Systems

If a system is conservative, the work done by the forces can be calculated from the potential energy P. We define the change in potential energy during a small displacement as the negative of the work done by the forces of the system during the displacement. Because $Q_1 \delta q_1 + Q_2 \delta q_2$ is the work done by the forces, we have

$$\delta P = -Q_1 \delta q_1 - Q_2 \delta q_2. \tag{A.24}$$

We have emphasized that q_1 and q_2 are independent and, hence, can be varied arbitrarily. If $\delta q_2 = 0$, we have $\delta P = -Q_1 \delta q_1$, so that

$$Q_1 = -\frac{\partial P}{\partial q_1}. \tag{A.25}$$

Similarly, it can be seen that

$$Q_2 = -\frac{\partial P}{\partial q_2}. \tag{A.26}$$

Replacing Q_1 and Q_2 in Eqs. (A.23) by these expressions we have

$$\begin{aligned}
\frac{d}{dt}\left(\frac{\partial K}{\partial \dot{q}_1}\right) - \frac{\partial K}{\partial q_1} + \frac{\partial P}{\partial q_1} &= 0, \\
\frac{d}{dt}\left(\frac{\partial K}{\partial \dot{q}_2}\right) - \frac{\partial K}{\partial q_2} + \frac{\partial P}{\partial q_2} &= 0.
\end{aligned} \tag{A.27}$$

These are Lagrange's equations of motion for a conservative system. As before, they hold for systems of any finite number of degrees of freedom.

Example 4

Find the equations of motion of a particle of weight W moving in space under the force of gravity.

Solution: We need three coordinates to describe the position of the particle and can therefore take x, y, z as our generalized coordinates. Taking x and y in the horizontal place and z vertically upward with the origin at the earth's surface and taking the origin as the zero position for potential energy, one obtains

$$K = \frac{W}{2g}(\dot{x}^2 + \dot{y}^2 + \dot{z}^2), \qquad P = Wz,$$

$$\frac{\partial K}{\partial \dot{x}} = \frac{W}{g}\dot{x}, \qquad \frac{\partial K}{\partial \dot{y}} = \frac{W}{g}\dot{y}, \qquad \frac{\partial K}{\partial \dot{z}} = \frac{W}{g}\dot{z}, \qquad \frac{\partial K}{\partial x} = \frac{\partial K}{\partial y} = \frac{\partial K}{\partial z} = 0,$$

$$\frac{d}{dt}\left(\frac{\partial K}{\partial \dot{x}}\right) = \frac{W}{g}\ddot{x}, \qquad \frac{d}{dt}\left(\frac{\partial K}{\partial \dot{y}}\right) = \frac{W}{g}\ddot{y}, \qquad \frac{d}{dt}\left(\frac{\partial K}{\partial \dot{z}},\right) = \frac{W}{g}\ddot{z},$$

$$\frac{\partial P}{\partial x} = \frac{\partial P}{\partial y} = 0, \qquad \frac{\partial P}{\partial z} = W. \tag{A.28}$$

Hence, Lagrange's equation, Eq. (A.27), gives

$$\frac{W}{g}\ddot{x} = 0, \qquad \frac{W}{g}\ddot{y} = 0, \qquad \frac{W}{g}\ddot{z} + W = 0. \tag{A.29}$$

Of course, these equations are more easily obtainable by the direct application

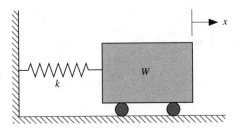

Figure A.1 Schematic for the mechanical system of Example 4.

of Newton's Second Law. This example merely illustrates the application of Lagrange's equations for a familiar problem.

Example 5

Find the equation of motion of the sprung weight W sliding on a smooth horizontal plane (see Fig. A.1).

Solution: We may take x as the generalized coordinate and measure it from the equilibrium position. Then

$$K = \frac{W}{2g}\dot{x}^2, \qquad P = \frac{k}{2}x^2,$$

$$\frac{\partial K}{\partial \dot{x}} = \frac{W}{g}\dot{x}, \qquad \frac{d}{dt}\left(\frac{\partial K}{\partial \dot{x}}\right) = \frac{W}{g}\ddot{x}, \qquad (A.30)$$

$$\frac{\partial K}{\partial x} = 0, \qquad \frac{\partial P}{\partial x} = kx.$$

Lagrange's equation, Eq. (A.27), gives

$$\frac{W}{g}\ddot{x} + kx = 0 \qquad (A.31)$$

as the equation of motion.

Example 6

Obtain the equations of motion for the system shown in Fig. A.2. The bar is weightless.

Solution: The coordinates x_1 and x_2 can be taken as generalized coordinates. Take as the zero datum the configuration for which the bar is horizontal and the

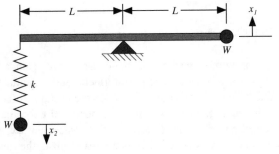

Figure A.2 Schematic for mechanical system of Example 6.

spring is unstretched. Then

$$K = \frac{W}{2g}\dot{x}_1^2 + \frac{W}{2g}\dot{x}_2^2, \qquad P = Wx_1 - Wx_2 + \frac{1}{2}k(x_2 - x_1)^2,$$

$$\frac{\partial K}{\partial \dot{x}_1} = \frac{W}{g}\dot{x}_1, \qquad \frac{\partial K}{\partial \dot{x}_2} = \frac{W}{g}\dot{x}_2, \qquad \frac{\partial K}{\partial x_1} = \frac{\partial K}{\partial x_2} = 0,$$

$$\frac{d}{dt}\left(\frac{\partial K}{\partial \dot{x}_1}\right) = \frac{W}{g}\ddot{x}_1, \qquad \frac{d}{dt}\left(\frac{\partial K}{\partial \dot{x}_2}\right) = \frac{W}{g}\ddot{x}_2, \qquad (A.32)$$

$$\frac{\partial P}{\partial x_1} = W - k(x_2 - x_1), \qquad \frac{\partial P}{\partial x_2} = -W + k(x_2 - x_1).$$

The Lagrange equations are

$$\frac{W}{g}\ddot{x}_1 + W - k(x_2 - x_1) = 0,$$

$$\frac{W}{g}\ddot{x}_2 - W + k(x_2 - x_1) = 0. \qquad (A.33)$$

This is an example of a two-degree-of-freedom, conservative system.

A.6   Lagrange's Equations for Nonconservative Systems

If the system is nonconservative, then, in general, there will be some forces (conservative) that are derivable from a potential function, $P(q_1, q_2, \ldots)$ and some forces (nonconservative) that are not. Those forces for which a potential function does not exist must be introduced by first determining their virtual work. The coefficient of the virtual displacement δq_i in the virtual work expression is the generalized force, here denoted by Q_i ($i = 1, 2, \ldots$). In this instance it is convenient to introduce what is called the Lagrangean as

$$L = K - P \qquad (A.34)$$

and write the general form of Lagrange's equations as

$$\frac{d}{dt}\left(\frac{\partial L}{\partial \dot{q}_i}\right) - \frac{\partial L}{\partial q_i} = Q_i \qquad (i = 1, 2, \ldots, n). \qquad (A.35)$$

Example 7

Rework Example 5 with a dashpot of constant c connected in parallel with the spring.

Solution: The system with a dashpot is nonconservative. Hence, we use Lagrange's equations in form of Eq. (A.35). The kinetic and potential energies are the same as in Example 5. To calculate the Q for the dashpot force, use the definition that Q is the coefficient by which the generalized coordinates must be multiplied to obtain the work done. In any small displacement δx, the work done by the dashpot force $-c\dot{x}$ is $-c\dot{x}\,\delta x$. Hence, $-c\dot{x}$ itself is the generalized force associated with the dashpot. The Lagrangean is

$$L = \frac{W\dot{x}^2}{2g} - \frac{kx^2}{2}$$

(A.36)

and

$$Q = -c\dot{x}.$$

(A.37)

Lagrange's equation then becomes

$$\frac{W}{g}\ddot{x} + kx = -c\dot{x}.$$

(A.38)

References

Structural Dynamics

[1] Bisplinghoff, R. L., H. Ashley, and R. L. Halfman, *Aeroelasticity*, Addison-Wesley Publishing Co., Inc., 1955.

[2] Chen, Y., *Vibrations: Theoretical Methods*, Addison-Wesley Publishing Co., Inc., 1966.

[3] Fry'ba, L., *Vibration and Solids and Structures under Moving Loads*, Noordhoff International Publishing, 1972.

[4] Hodges, D. H., *Nonlinear Composite Beam Theory*, American Institute of Aeronautics and Astronautics, 2006.

[5] Hurty, W. C., and M. F. Rubenstein, *Dynamics of Structures*, Prentice-Hall, Inc., 1964.

[6] Kalnins, A., and C. L. Dym, *Vibration: Beams, Plates and Shells*, Dowden, Hutchinson and Ross, Inc., 1976.

[7] Lamb, H., *The Dynamical Theory of Sound*, 2nd ed., 1925 (reprinted by Dover).

[8] Leipholz, H., *Direct Variational Methods and Eigenvalue Problems in Engineering*, Noordhoff International Publishing, 1977.

[9] Lin, Y. K., *Probabilistic Theory of Structural Dynamics*, McGraw-Hill, Inc., 1967; reprinted by Robert E. Krieger Publishing Co., 1976.

[10] Meirovitch, L., *Computational Methods in Structural Dynamics*, Sijthoff and Noordhoff, 1980.

[11] Meirovitch, L., *Principles and Techniques of Vibrations*, Prentice-Hall, Inc., 1997.

[12] Pestel, E. C., and F. A. Leckie, *Matrix Methods in Elastomechanics*, McGraw-Hill Book Co., Inc., 1963.

[13] Rayleigh, J. W. S., *The Theory of Sound*, Vol. I, 1877, The MacMillan Co., 1945 (reprinted by Dover).

[14] Scanlan, R. H., and R. Rosenbaum, *Introduction to the Study of Aircraft Vibration and Flutter*, The MacMillan Co., 1951 (reprinted by Dover).

[15] Snowdon, J. C., *Vibration and Shock in Damped Mechanical Systems*, John Wiley and Sons, 1968.

[16] Steidel, R. F., Jr., *An Introduction to Mechanical Vibrations*, 2nd ed., John Wiley and Sons, 1979.

[17] Thomson, W. T., and M. D. Dahleh, *Theory of Vibration with Applications*, 5th ed., Prentice-Hall, Inc., 1998.

[18] Timoshenko, S., and D. H. Young, *Vibration Problems in Engineering*, 3rd ed., Van Nostrand Reinhold Co., 1955.

[19] Tse, F. S., I. E. Morse, and R. T. Hinkle, *Mechanical Vibrations: Theory and Applications*, 2nd ed., Allyn and Bacon, Inc., 1978.

Aeroelasticity

[1] Abramson, H. N., *An Introduction to the Dynamics of Airplanes*, Ronald Press Co., 1958 (reprinted by Dover).

[2] Anon., "Manual on Aeroelasticity," six volumes by NATO Advisory Group for Aeronautical Research and Development, 1956–1970.

[3] Bisplinghoff, R. L., and H. Ashley, *Principles of Aeroelasticity*, John Wiley and Sons, Inc., 1962 (reprinted by Dover).

[4] Bisplinghoff, R. L., H. Ashley, and R. L. Halfman, *Aeroelasticity*, Addison-Wesley Publishing Co., Inc., 1955.

[5] Collar, A. R., "The First Fifty Years of Aeroelasticity," *Aerospace*, Vol. 5 (Paper No. 545), February 1978, pp. 12–20.

164

[6] Diederich, F. W., and B. Budiansky, "Divergence of Swept Wings," NACA TN 1680, 1948.

[7] Dowell, E. H., *Aeroelasticity of Plates and Shells*, Noordhoff International Publishing, 1975.

[8] Dowell, E. H., E. F. Crawley, H. C. Curtiss Jr., D. A. Peters, R. H. Scanlan, and F. Sisto, *A Modern Course in Aeroelasticity*, 3rd ed., Kluwer Academic Publishers, 1995.

[9] Freberg, C. R., and E. N. Kemler, *Aircraft Vibration and Flutter*, John Wiley and Sons, Inc., 1944.

[10] Fung, Y. C., *An Introduction to the Theory of Aeroelasticity*, John Wiley and Sons, Inc., 1955 (reprinted by Dover).

[11] Garrick, I. E., and W. H. Reed III, "Historical Development of Aircraft Flutter," *Journal of Aircraft*, Vol. 18, No. 11, Nov. 1981, pp. 897–912.

[12] Hassig, H. J., "An Approximate True Damping Solution of the Flutter Equation by Determinant Iteration," *Journal of Aircraft*, Vol. 8, No. 11, Nov. 1971, pp. 885–889.

[13] Irwin, C. A. K., and P. R. Guyett, "The Subcritical Response and Flutter of a Swept Wing Model," Tech. Rep. 65186, Aug. 1965, Royal Aircraft Establishment, Farnborough, UK.

[14] Peters, D. A., S. Karunamoorthy, and W.-M. Cao, "Finite State Induced Flow Models; Part I: Two-Dimensional Thin Airfoil," *Journal of Aircraft*, Vol. 32, No. 2, Mar.–Apr. 1995, pp. 313–322.

[15] Scanlan, R. H., and R. Rosenbaum, "Outline of an Acceptable Method of Vibration and Flutter Analysis for a Conventional Airplane," *CAA Aviation Safety Release 302*, Oct. 1948.

[16] Scanlan, R. H., and R. Rosenbaum, *Introduction to the Study of Aircraft Vibration and Flutter*, The MacMillan Co., 1951 (reprinted by Dover).

[17] Theodorsen, T., "General Theory of Aerodynamic Instability and the Mechanism of Flutter," NACA TR 496, 1934.

[18] Weisshaar, T. A., "Divergence of Forward Swept Composite Wings," *Journal of Aircraft*, Vol. 17, June 1980, pp. 442–448.

Index

167